PETERSON FIELD GUIDE TO

FINDING
MAMMALS
in North America

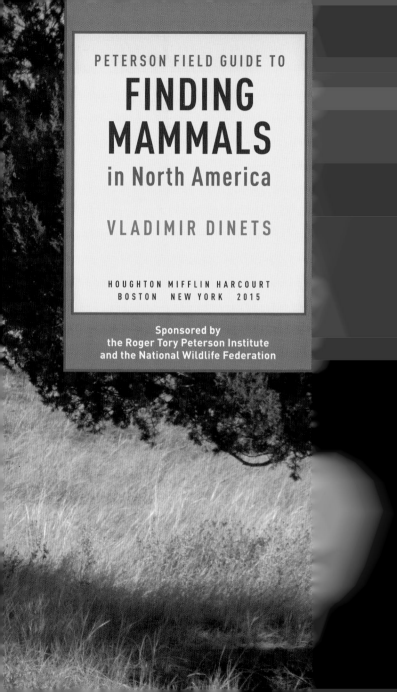

PETERSON FIELD GUIDE TO

FINDING MAMMALS

in North America

VLADIMIR DINETS

HOUGHTON MIFFLIN HARCOURT
BOSTON   NEW YORK   2015

Sponsored by
the Roger Tory Peterson Institute
and the National Wildlife Federation

For information about permission to reproduce selections from this book,
write to Permissions, Houghton Mifflin Harcourt Publishing Company,
215 Park Avenue South, New York, New York 10003.

www.hmhco.com

PETERSON FIELD GUIDES and PETERSON FIELD GUIDE SERIES are
registered trademarks of Houghton Mifflin Harcourt Publishing Company.

Library of Congress Cataloging-in-Publication Data is available.
ISBN 978-0-544-37327-3

Printed in China

SCP 10 9 8 7 6 5 4 3 2 1

To Nastia, my beloved wife, best friend,
heavenly muse, and motivational speaker.

The legacy of America's greatest naturalist and creator of the field guide series, Roger Tory Peterson, is kept alive through the dedicated work of the Roger Tory Peterson Institute of Natural History (RTPI). Established in 1985, RTPI is located in Peterson's hometown of Jamestown, New York, near Chautauqua Institution in the southwestern part of the state.

Today RTPI is a national center for nature education that maintains, shares, and interprets Peterson's extraordinary archive of writings, art, and photography. The Institute, housed in a landmark building by world-class architect Robert A. M. Stern, continues to transmit Peterson's zest for teaching about the natural world through leadership programs in teacher development as well as outstanding exhibits of contemporary nature art, natural history, and the Peterson Collection.

Your participation as a steward of the Peterson Collection and supporter of the Peterson legacy is needed. Please consider joining RTPI at an introductory rate of 50 percent of the regular membership fee for the first year. Simply call RTPI's membership department at (800) 758-6841 ext. 226, or email *membership@rtpi.org* to take advantage of this special membership offered to purchasers of this book. For more information, please visit the Peterson Institute in person or virtually at *www.rtpi.org*.

CONTENTS

LIST OF ABBREVIATIONS

CA – Conservation Area
CP – County Park
ER – Ecological Reserve
GS – Game Sanctuary
I - (followed by a number) –
    Interstate Highway
MNM – Marine National Monument
MPA – Marine Protected Area
NA – Natural Area
NCA – National Conservation Area
NF – National Forest
NG – National Grassland
NHP – National Historic Park
NLS – National Lakeshore
NML – National Memorial
NMS – National Marine Sanctuary
NMT – National Monument
NP – National Park
NPR – National Preserve
NR – National Reserve
NRA – National Recreation Area
NSS – National Seashore

NWR – National Wildlife Refuge
PP – Provincial Park
PR – Preserve
PWP – Provincial Wildlife Park
RA – Recreation Area
RNA – Research Natural Area
SF – State Forest
SGS – State Game Sanctuary
SHP – State Historic Park
SHS – State Historic Site
SNA – State Natural Area
SP – State Park
SPR – State Preserve
SR – State Reserve
SRA – State Recreation Area
SWA – State Wildlife Area
WA – Wilderness Area
WCA – Wetlands Conservation Area
WLA – Wildlife Area
WMA – Wildlife Management Area
WR –Wildlife Refuge (or Reserve)
WS – Wildlife Sanctuary

Two-letter postal codes are used for U.S. states and
Canadian provinces and territories, as follows:

AB	Alberta		ND	North Dakota
AL	Alabama		NE	Nebraska
AK	Alaska		NH	New Hampshire
AR	Arkansas		NJ	New Jersey
AZ	Arizona		NL	Newfoundland & Labrador
BC	British Columbia		NM	New Mexico
CA	California		NS	Nova Scotia
CO	Colorado		NT	Northwest Territories
CT	Connecticut		NU	Nunavut
DE	Delaware		NV	Nevada
DC	District of Columbia		NY	New York
FL	Florida		OH	Ohio
GA	Georgia		OK	Oklahoma
HI	Hawaii		ON	Ontario
IA	Iowa		OR	Oregon
ID	Idaho		PA	Pennsylvania
IL	Illinois		PE	Prince Edward Island
IN	Indiana		QC	Québec
KS	Kansas		RI	Rhode Island
KY	Kentucky		SC	South Carolina
LA	Louisiana		SD	South Dakota
MA	Massachusetts		SK	Saskatchewan
MB	Manitoba		TN	Tennessee
MD	Maryland		TX	Texas
ME	Maine		UT	Utah
MI	Michigan		VA	Virginia
MN	Minnesota		VT	Vermont
MO	Missouri		WA	Washington
MS	Mississippi		WI	Wisconsin
MT	Montana		WV	West Virginia
NB	New Brunswick		WY	Wyoming
NC	North Carolina		YT	Yukon Territory

ACKNOWLEDGMENTS

I thank the employees of all protected natural areas mentioned in the book for their work and their help. Special thanks to Cheryl Antonucci, Steve Babbs, Karen Baker, Nancy Black, Paul Carter, Morgan Churchill, Sharon Collinge, Alan Dahl, Steve Davis, John Dixon, Joe Dlugo, Suzi Eszterhus, Scott Flamand, John Fox, Steven Green, Curtis Hart, Charles Hood, Mark Hows, Scott Knapp, Alex Kogan, Olga and Michael Kosoy, Boris Krasnov, Paul Lehman, Steve Linsley, Stefan Lithner, Natalia Loginova, Jenella Loye, Matt Miller, Steve Morgan, Heidi Perryman, Paul Polechla, Luke Powell, Chris Ray, Mark Richardson, Mike Richardson, Don Robertson, Venkat Sankar, Gina Sheridan, Coke and Som Smith, Steve-Anyon Smith, Dale Steel, Philip Telfer, Richard Ternullo, Anastasiia Tsvietkova, Richard Webb, James White Hare, and particularly Jon Hall and Fiona Reid for help with field research and information gathering.

Also, my thanks to the team at Houghton Mifflin Harcourt for making this book as good as it is, and to my agent Regina Ryan for bringing us all together. Some photos were coauthored with Anastasiia Tsvietkova.

American Badger can be difficult to find, but once found, it is often amazingly tame and can be observed up close for hours.

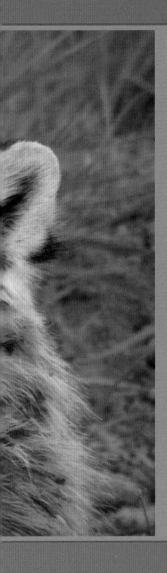

PART I.
THE ART OF
MAMMAL
WATCHING

Why Watch Mammals?

One hundred years ago, bird watching was a little-known hobby, practiced by a small number of people considered by many to be what today are called "geeks." There were no pocket-size field guides, and many bird species were believed to be indistinguishable in the wild. Only professional ornithologists with access to large museum collections were trusted to identify birds. But things gradually changed. Now bird watching (or birding, as it is increasingly called) is the fastest-growing kind of tourism.

Birders number in the millions in North America and Europe, and you can find lively birding communities in places like Moscow, Delhi, Rio de Janeiro, and Cape Town. Bird watching has largely replaced more destructive activities such as collecting bird eggs and mounted birds. It creates a wealth of useful data for ornithologists, strong incentives for bird conservation, and booming markets for bird books, which are published in multitudes every year.

Mammal watching today is exactly where bird watching was a century ago. There are few good field guides for mammals. Many species are routinely claimed to be identifiable only if caught, or only in a genetics lab. Only in the last few years has the situation begun to change, and now the popularity of mammal watching is growing exponentially as people discover that it can be at least as exciting and interesting as bird watching.

Mammals are more diverse than birds in size, body structure, and lifestyle. There are aquatic and flying mammals, but there are no subterranean or completely aquatic birds. So mammal watching is a

LEFT: *Once believed to be extinct in the wild, Black-footed Ferret can now be seen at reintroduction sites, particularly in the Aubrey Valley of Arizona. Here is a female carrying her baby to a new den.* ABOVE: *Some North American mammals are still waiting to be discovered. This is the mysterious cottontail from the Cumberland Plateau, probably an undescribed species.*

ABOVE: *Once you get a close view of the cute little face of an Eastern Pipistrelle, you will never be afraid of bats again.*

RIGHT: *White-nosed Coatis are diurnal, social, and very inquisitive. They are easy to find in Arizona if you know where to look.*

more diverse experience. But that's not the only reason finding mammals is so challenging and rewarding.

Birds are usually the easiest animals to see. After just a year or two of bird watching, most people run out of new species to look for in their state or province. All regularly occurring species in North America can be seen within four or five years if you don't mind spending a few thousand dollars on airfare and gasoline. In fact, some people manage to find them all in one year.

Mammals tend to be more cryptic. I've been looking for them in

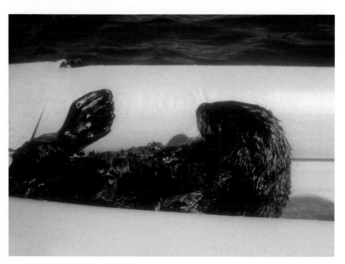

A wild Sea Otter taking a ride in my kayak.

North America for 17 years, and there are still a few species I haven't seen. Switching from bird to mammal watching takes you to a new level of intimacy with the wilderness. You have to be much more observant, attentive, and sensitive to subtle clues. You have to feel as at home in the woods or desert at night as you do during the day. You need to learn the alphabet of animal tracks, the language of rustling dry leaves, the scents of forest trails, the interplay of ocean currents. You must learn to move as quietly as a shadow, to wait for hours without moving, to turn into snags, tussocks, clumps of seaweed. You don't have to be particularly athletic (the first naturalist to observe Evoron Vole in the wild did it from a wheelchair when he was well over 70), but training your mind takes time.

Once you master the required skills, you can experience nature in a different way. Every broken twig, every footprint in the snow, every little burrow in the sand tells you a story. You witness scenes that few, if any, mortals have been privileged to see since the days of the last great Native American hunters. I have watched Jaguars playing in the moonlight, Killer Whales hunting in towering kelp forests, a vanishingly rare Black-footed Ferret carrying her naked pups to a new den. A baby Striped Skunk played with my toes, a Gray Whale calf let me rub its nose, a walnut-sized Silky Pocket Mouse ate from my hand, a Sea Otter left her cub in my kayak while diving for sea urchins. You will never forget the day when you see your first Star-nosed Mole in New England woods, or hear your first Gray Wolf howl in a tundra blizzard, or pet a Harp Seal cub on Canadian ice floes.

Compared to birds, mammals are still little known. New species are being described in North America almost every year, and these include not just small creatures, but even whales. Some of these descriptions are unlikely to be accepted by the scientific community in the long run, but others are well substantiated. As for the species already named, the natural history of many of them is poorly understood. Lives of many shrews, rodents, bats, and marine mammals remain mysterious. Any good observer has a chance to make a zoological discovery by studying mammals. And, unlike hunters, you will not destroy the wonders you discover—at least as long as you are careful and gentle.

How to Find Them

To an inexperienced person, forests and grasslands often seem devoid of mammals, except for a handful of conspicuous species such as squirrels and deer. But in reality, most North American environments are packed with mammals; you just have to know when and where to look, and how to avoid scaring them away.

Forest animals are usually most active for a few hours after sunset and around sunrise, but in places with little or no human presence they will start appearing in late afternoon. Try to spend your evenings, early nights, and early mornings outdoors. In campgrounds and other places with lots of people, the best time is after midnight, when humans go to sleep and animals can finally come out. Desert animals tend to be even more strictly nocturnal, but can be active during the day in winter, or visit watering holes in late morning or afternoon in hot weather.

Many small mammals are much less active on moonlit nights, while others avoid being outside when it rains. Dry weather makes it easy to locate small forest creatures by the rustling sound they make in fallen leaves. But if you are looking for larger animals, morning dew, wind, light rain, or a fast stream nearby will make it easier for you to sneak up on them.

Learning to walk quietly is easier than most people think. Slow down the movement of your foot just before it touches the ground; don't raise your feet too high; when you are close to the animal, roll

Observing animal tracks can be as much fun as watching the animals themselves. These are week-old Arctic Fox tracks sculpted by the wind.

your foot from heel to toes using the outside edge of your sole. If there are a lot of dry leaves, put your foot down flat, as if walking on ice.

Clothing is the main source of noise while walking; keep it to a minimum (unfortunately, North America is not the best place to walk around naked) and avoid noisy jackets. I usually wear sandals with soft soles and covered toes, except when it's too cold.

Speaking of clothing: unless it's hunting season and you have to wear orange, try to wear camouflage, or at least clothes that more or less match the background colors. Many mammals are fully or partially colorblind, but some aren't; besides, birds see color very well, and mammals often understand their alarm calls.

It's better to walk alone; two people is the maximum suggested group size. It's not only because most people just can't keep from talking, but also because your attention inevitably shifts from the outside world to your companions. If you have to walk with more people, try walking a few steps ahead or behind; that way you can mentally block some of the noise.

As soon as you walk into the forest far enough to escape road sounds, stop for a few minutes and listen. Tune your hearing to leaves rustling, birds singing, trees moving. Imagine yourself a spider in the center of a web, with subtle signals streaming in from all directions. If you walk for a long time, it's a good idea to stop for a few seconds now and then and listen, or your senses will gradually relax and you'll start missing things.

If you hear a noise, stop and try to find the source. Was it a drying leaf, a falling twig, a running shrew? Gradually you will develop a subconscious library of sounds, and will be able to tell those noises apart intuitively. In a few months you'll be able to tell a mouse from a

American Beavers are a lot of fun to watch, but you have to be very quiet and patient.

lizard by ear from a hundred feet away. Once you find an interesting animal, freeze . . . and it might approach you and even sniff your toes.

Small mammals (roosting bats, baby mice in nests, fighting shrews) often make very high-pitched noises. Our sensitivity to such sounds declines with age, so if you have children it's better to start bringing them along as soon as they learn to behave properly in the woods. If you don't have children or they are not outdoor types, you can train your dog to assist you, but some of the best mammal-watching sites don't allow pets.

I usually try to combine active and passive search. Walk around for half an hour, sit near a trail or a stream for a while, then walk again. Always try to walk a trail at least once in daylight before going there at night. All-night watches make sense only in really good locations, such as at mineral licks; among fruiting trees; on logs forming natural bridges across streams; around recently fallen large trees; at porous limestone outcrops, desert springs, trail junctions, and bottoms of narrow canyons; and on trails following rivers, cutting across river bends, or connecting two forest patches. Places where two different habitats come together, such as wood margins, marsh edges, and areas where cold and warm currents mix at sea, are also good.

A tent or a car can make a perfect blind. If you have to sit in the open, don't forget mosquito repellent. It will mask your scent a bit and prevent you from scaring away animals by scratching bites and swatting mosquitoes.

Don't limit your walks to trails. Narrow forest roads can also be good for night hikes or drives, especially if there is no traffic. Some places have paved trails or boardwalks; these are perfect for *spotlighting* (looking for animals at night by walking around with a flashlight) because you can move quietly without watching your step too much (just try to avoid stepping on snakes).

I like walking along forest streams, especially during droughts. The shores are often overgrown, but you can wade along or swim. Be careful doing this in the Southeast, where alligators and nutria itch worms occur. An inflatable kayak is excellent for exploring narrow waterways: it's portable, very quiet, and has shallow draft.

It is good to have binoculars with zoom. They should have a wide-view angle (to follow arboreal rodents in vine tangles, for example), but enough magnification to see bats roosting 50 to 100 feet above the ground. Good light-gathering capacity is also important. At night, in large caves, and in tall hollow conifers such as redwoods you'll have to use binoculars and a flashlight together; practice this in advance.

A reliable light source is the most important piece of equipment. In open grasslands and deserts you might need a huge spotlight, but for forests and most caves, smaller LED flashlights are sufficient. For night walks I usually use a small headlamp (which I keep turned on most of the time, except on moonlit roads) and a larger flashlight (which I only use occasionally). The small headlamp allows me to see the eyeshine of animals without scaring them away; the larger flashlight can be used to get a better look at an animal you've found, or to search through tree crowns around forest clearings. Many species, particularly bats, are much less disturbed by red light. Good headlamps have red light mode; you can also use red plastic film to make removable red filters for larger flashlights, or use red camera filters. In addition to minimizing disturbance, red light doesn't attract insects (very important if you use a headlamp). Headlamps take some getting used to, and don't work well in dense fog or drizzle. Using rechargeable batteries eventually saves you a lot of money.

Contrary to many cartoons and horror movies, animal eyes do not glow in the dark; they reflect only bright light, mostly in the same direction that it comes from. When spotlighting, try to hold the flashlight in such a way that your line of sight will be parallel to, and close to, the light beam—that way you'll have the best chance of seeing the reflection of the light in animals' eyes. Headlamps are particularly good for this.

I always enjoy exploring caves and abandoned mines, but many people find them dangerous (see? I warned you, so I'm not responsible if something goes wrong). **Never** enter any caves unless you have three flashlights with fully charged batteries, and a few spare batteries in securely closed pockets. This simple rule has saved my life a couple times. (By the way, it's also a good idea to have a spare flashlight during night walks.) Even in an electrified showcave a flashlight might come in handy. Wear a hard hat, and if you see a box of ancient-looking dynamite sticks in an old mine, don't kick it around (people have died that way). Small and large caves often have different inhabitants, even if they are located in the same area. Don't forget to thoroughly wash your clothes and shoes (preferably with bleach) after visiting caves. If you move between states, it's a good idea to buy new caving clothes and shoes to make absolutely sure you don't spread any pathogens.

Some species such as tree voles spend their entire lives in the forest canopy. Trying to see them from below can cause severe neck pain. There are many ways to get closer to the canopy. You can find a steep slope with good canopy lookouts, or learn tree-climbing techniques. Local tree-trimming companies sometimes have tree-climbing gear for rent.

Old bird nests, tree hollows, and various burrows are often used by small mammals. Look also under loose tree bark, in rock niches and cracks, and in piles of dead cacti or agave leaves in the desert. A

Hibernating bats are often seen in winter during commercial cave tours.

small mirror attached to a piece of wire can be used to look inside woodpecker holes, nest boxes, and other narrow spaces. If a hole in a tree is too high, try scratching the trunk—sometimes the inhabitants will peek out. Be careful: snakes, bees, and wasps often live in tree hollows. Carry a small flashlight with you on daytime hikes, too.

Many small animals hide under rocks and logs. Not all logs are good, though. An ideal log has been in place for a long time, but is not yet too rotten or overgrown (only some shrews prefer very rotten logs). You can see some small mammals by placing thick plywood sheets in the forest or in a fallow field, and looking underneath them about once a week (best at night).

Some rubber toys make sounds similar to the calls of small rodents. Such toys can be used to attract small carnivores. There are also special "mouse squeakers," sold in hunters' stores and online; the cheapest ones are usually the best. Hunters use commercially available tapes with distress calls of baby rabbits, mating calls of cougars, and other sounds potentially attractive to wildlife. I've tried them a few times but never had much success. However, once you find a weasel, a marten, or a fox, a good mouse squeaker can be really useful for getting a better look at it. "Pishing" sounds that birders make to attract birds also work sometimes, on carnivores as well as on chipmunks.

Larger mammals are usually less common within a few kilometers of towns, but there are many exceptions. Relatively tame animals will visit ranger stations, campgrounds, and ancient ruins. Many carnivores, rodents, and Virginia Opossum regularly visit garbage dumps. You have to be particularly careful around "garbage bears" and other large mammals that have learned to associate people with food. Don't get too close; it's always more interesting to observe natural behavior than to cause the animal to flee or attack you.

Bird feeders occasionally attract mammals during the day and at night. In parts of the Southwest, nectar-feeding bats often visit hummingbird feeders. Some people use baiting extensively; they leave piles of cat food, meat scraps, or peanuts in the forest and watch them from nearby blinds. Use this method very sparingly; it's important to make sure that the animals don't learn to associate food with people. Becoming too tame might cost them dearly, and large mammals can become dangerous.

Many rodents can be easily caught with live traps (Sherman traps are the kind most commonly used in North America). It is possible to

Swamp Rabbit is crepuscular, which means it is active mostly at dusk, and hides at other times of day.

see all North American mammals without trapping, but if you must use this method, there are a few important rules to follow. Don't use these traps during the day, or you'll end up killing lots of birds and lizards. If the traps are not fully shaded, remove or close them before the sun rises. Always put in enough bait to last the caught animal all night. If the temperature might fall below 50°F (15°C), put some natural cotton in the trap so that the animal can make a warm nest. Mark trap locations with colored tape so that you can find them the next morning. In most regions, a mixture of oats and peanut butter is used for bait, but in places with lots of fire ants (i.e., all of the Southeast and parts of California) you should use dry cat food, sunflower seeds, or cracked corn, and spray ant repellent around each trap (or simply refrain from trapping in summer). In areas with lots of shrews, such as northern swamps and meadows, check the traps around midnight as well as at dawn. It is better to release the animals in shaded, even dark, places or at night—they might behave in a more relaxed way, and even allow you to observe them for a while. To avoid spreading disease, soak the traps in hot water with bleach after every capture and before moving them to a new area, and always wear rubber gloves and a face mask when trapping rodents.

Learning the art of tracking is extremely helpful (a couple of good manuals are listed in Part IV). It's seldom possible to simply follow the tracks until you catch up with the animal (although sometimes this method does work, especially for small mammals in sandy areas). But looking at tracks will often tell you which areas are used by animals most regularly. Many species, ranging from shrews to bears and bison, create networks of trails (rodent trails are often called *runways*) and can sometimes be seen by waiting near those routes. Also, reading animal tracks and signs provides you with deep understanding of the everyday life of the wilderness that can't be substituted for by any books.

There are countless other methods, and you will come up with some of your own eventually. Some tips for finding particular kinds of animals are given in Part III. Try to be outdoors as much as you can, and always keep looking and listening. You can spend a whole night on a trail and see nothing, but the next night the same trail may be full of wildlife.

Good hunting.

Before You Leave Home

North America is a continent of rules and regulations. You have to know them well before you go anywhere—if you trespass on unmarked private property, stay in nature reserves after hours, or walk into a forest without purchasing a permit, you risk being fined, arrested, or even shot. Entering most, but not all, national parks and many state parks requires paying an entry fee. National forests and wildlife management areas are usually free, but there are a growing number of exceptions. In the U.S. you can save money if you buy an annual pass for all national parks, which covers everybody in your car. A more expensive version also covers national wildlife refuges, but those are often free anyway or closed to visitors.

Virtually every protected natural area in North America has a website. These range from simple pages for county parks and wildlife management areas to online encyclopedias that some national parks have. Always check them before visiting, not just for general information, but also for practicalities like hours, road closures, permits required, and so on. You can also email requests for particular information, or just show up at park offices, visitor centers, and ranger stations and talk to the personnel. Once they understand what you are interested in, they might be able to share some great tips about local wildlife.

Some populations of Desert Bighorn are listed as endangered, so they should be observed and photographed only from afar, especially during the lambing season.

A very important issue for mammal watchers is nighttime access. Most national wildlife refuges have switched to daylight-only access in recent years, but you can try applying for a "special use" permit. National parks, forests, and grasslands are usually open at night, but state, provincial, county, and city parks may be closed, or open only for those who stay at campgrounds inside; in most parks you are not allowed to camp except in designated places or to sleep in your car. Campgrounds seldom qualify as "staying outdoors," as they tend to be huge, overcrowded, and noisy, and the worst ones are very citylike, but some of them attract interesting animals. Note that for many residents of North America "camping" doesn't mean sleeping in a tent; these people "camp" by driving a house-sized motor home to one of those campgrounds and watching TV inside most of the night. Overcrowding is a huge problem in many parks in the U.S. and in some parks in Canada, but usually all you have to do to find solitude is walk away from the pavement for five minutes. Many people in developed countries dislike walking; their general idea of communicating with nature is driving around for days, hardly ever setting foot on soft ground (I have to admit I've caught myself trying to do this a few times). Most campers are also strictly diurnal, so the parts of parks away from campgrounds tend to be completely deserted at night. An exception is Anhinga Trail in the Everglades, a popular place to spotlight for alligators.

Plains Bison look beautiful from almost any distance. It's always better to observe a naturally behaving animal from afar than to approach too closely and risk scaring or provoking it.

Spotlighting from cars is not allowed in many parks (notably in Yellowstone National Park), and rules regulating live-trapping can be sketchy or contradictory. Playback of animal sounds, baiting, and using scent lures (such as beaver spray and pheromone-based lures used by hunters) are generally prohibited in national parks. You can often obtain a permit, but that takes time, and you will most likely be asked to demonstrate the scientific merit of what you are planning to do.

If tightly regulated national and state parks seem too oppressive for you, try Bureau of Land Management lands; they are often as close to perfect freedom as one could wish for. Unfortunately, there are almost no BLM lands in the eastern U.S., and finding a good place for night walks might be difficult in some eastern states. In Canada there are plenty of unregulated forest tracts.

If you consider live-trapping, note that local laws and park rules often don't differentiate between catch-and-release and commercial trapping; always check with local rangers or wildlife authorities to avoid getting in trouble.

Trespassing on private land can result in arrest, particularly in the southeastern U.S. However, if you manage to locate the owner (which can take a few days or even months in some cases), it is usually possible to get access permission, unless the owner is a large

Brush Mouse. Wild rodents should never be handled without protective gloves, and in the West it's better to avoid handling them at all.

company with a macabre set of bureaucratic rules. Knowing basic Spanish will help you a lot in agricultural areas, and anywhere in southern Florida and the southwestern U.S. In Québec (except for the far north of the province, where people speak mostly English and Cree) it's a good idea to start conversations with "*Bonjour! Excusez-moi, je ne parle pas français*" (Hello! Excuse me, I don't speak French) . . . unless, of course, you do speak French. In Greenland a lot of people speak English; the local languages are Greenlandic Inuit and Danish, of which the latter is much easier to learn for English speakers.

Native American lands (officially called First Nations lands in Canada) have their own rules, sometimes bordering on extortion (for example, on Navajo Nation lands you are not allowed to walk off-road without a local guide, while in some New Mexican pueblos you

Whale watching is a lot of fun, but if you try it by yourself, make sure you know all the rules.

Harbor Seals can get used to people, but it's illegal to feed them.

can be fined $10,000 for using a camera). Consulting in advance with the tribal council might be a good idea. If you are planning to travel in areas with a largely Native population, consider learning at least the words for "hello" and "thank you" in the local language; this is helpful as well as polite. Visiting Native lands is often worth the trouble, since many tribes do a really good job of protecting local fauna.

You have to be particularly careful with federally or locally listed endangered species. In some cases simply watching them with a spotlight might be considered harassment and result in a severe penalty. Accidentally trapping a jumping mouse or a shrew of an endangered subspecies might also get you into a lot of trouble. It is a good idea to refrain from any invasive activities in places where endangered animals can occur.

In U.S. waters, the Marine Mammal Protection Act sets the limits on what is considered harassment; these rules are complicated and difficult to find. Generally you are not allowed to approach marine mammals (including sea otters and manatees) to less than 150 feet, touch them, feed them, or cause them to move away. It is, however, okay to wait for the animals to approach you, and in Florida you are allowed to gently touch playful manatees, but only with one hand at a time. In many areas these rules are routinely ignored by tourists, fishermen, and tour operators, but in some cases the restrictions have been interpreted very broadly and people have been prosecuted for the slightest infractions. Similar restrictions exist in Canada, although they don't apply to commercial seal hunters who kill thousands of Harp Seal pups every year.

Environmental rules and regulations might seem overcomplicated, but they more or less do the job. North America is the only part of the world (except the Antarctic) where no species of mammal has gone extinct in the last 50 years. Let's keep it that way.

Moose in the Colorado Rockies.

PART II.
THE BEST
LOCATIONS

North America north of Mexico has more than 420 species of native mammals, as well as a growing number of introduced ones. Of the native species, 64 are marine, and at least 167 are *endemic*, meaning that they do not naturally occur anywhere else (although some North American species have been introduced to other continents). A few Mexican mammals, such as Nine-banded Armadillo and probably some bats, have recently expanded their range into the U.S. With the ongoing climate change, this process can be expected to accelerate.

In the approximately 15,000 years (a highly controversial estimate) since human arrival to North America, the majority of large terrestrial mammals (over 50 species), at least one marine mammal (Steller's Sea Cow), and an unknown number of small mammals have gone extinct. The degree to which human hunting has caused their extinction is also highly controversial, but it is increasingly accepted that humans played a major role. Two more species (Sea Mink and Caribbean Monk Seal) and many subspecies have become extinct since the Europeans' arrival. A few subspecies, such as Plains Bison, Mexican Wolf, and "typical" Red Wolf, became extinct in the wild and have been reintroduced from captive populations. Black-footed Ferret was considered extinct in the wild, but recently a surviving wild population has been discovered, in addition to the growing number of reintroduced ones. Northern Elephant Seal was considered extinct from 1884 to 1912, and Guadalupe Fur Seal from 1928 to 1954, but subsequently small breeding populations were found and are now thriving. Based on the chronological patterns of discovering new species, it can be estimated that 5 to 25 species of mammals in North America remain undiscovered, and this number does not necessarily include only small animals.

California and Nevada

California and Nevada are a study in contrasts. California is terribly overcrowded; its major population centers are located in the worst imaginable places to build large cities; its environment has suffered gold rushes, timber booms, construction booms, and some of the most hideous abuses of natural water bodies ever committed by engineers. Still, it remains one of the world's top 10 destinations for naturalists, and harbors outstanding biological diversity, as well as some

LEFT: *Sierra Bighorn Sheep is one of the rarest mammals in North America.*

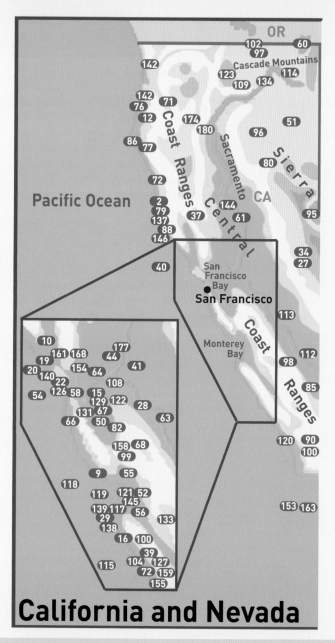

California and Nevada

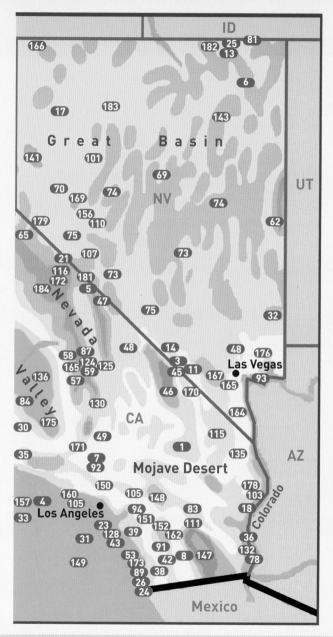

ID

166

182 25 81
13

6

17 183

143

Great   Basin

141 101

69

70 74 NV

169 74

156 74

110 62

179

65 75

21 107 73

116 181 73

172 5

184 47

75 32

Nevada

48 14

3

48 176

87 11 Las Vegas

58 124 45 167

165 59 125 46 170 165 93

57

130 164

84 CA 115

30 175

49 135 AZ

171

35 1

7

92 Mojave Desert 178

150 103

160 105 148 18

157 4 94 83

33 Los Angeles 151 152 111 36

23 39 162 132

31 128 91 78

43 42 8 147

53

149 173

89 38

26

24 Mexico

Valley

Colorado

California and Nevada

1. Afton Canyon NA (CA)
2. Albion (CA)
3. Amargosa Valley (NV)
4. Anacapa I. (CA)
5. Ancient Bristlecone Pine Forest (CA)
6. Angel Lake (NV)
7. Antelope Valley California Poppy Reserve (CA)
8. Anza-Borrego Desert SP (CA)
9. Año Nuevo SP (CA)
10. Armstrong Redwoods SR (CA)
11. Ash Meadows NWR (NV)
12. Azalea Reserve (CA)
13. Bear Creek Summit (NV)
14. Beatty (NV)
15. Berkeley (CA)
16. Big Sur (CA)
17. Black Rock Desert (NV)
18. Blythe (CA)
19. Bodega Bay (CA)
20. Bodega Submarine Canyons (CA)
21. Bodie SHP (CA)
22. Bolinas Lagoon (CA)
23. Bolsa Chica ER (CA)
24. Border Field SP (CA)
25. Bruneau Meadows (NV)
26. Cabrillo NMT (CA)
27. Calaveras Big Trees SP (CA)
28. Caswell Memorial SP (CA)
29. Carmel Bay (CA)
30. Carrizo Plain NMT (CA)
31. Catalina I. (CA)
32. Cathedral Gorge SP (NV)
33. Channel Is. NP (CA)
34. China Flat Campground (CA)
35. Chumash Painted Cave SHP (CA)
36. Cibola NWR (AZ/CA)
37. Clear Lake SP (CA)
38. Cleveland NF (CA)
39. Cone Peak (CA)
40. Cordell Bank NMS (CA)
41. Cosumnes River PR (CA)
42. Cuyamaca Rancho SP (CA)
43. Dana Pt. Headlands (CA)
44. Davis (CA)
45. Death Valley Jct. (CA)
46. Death Valley NP (CA/NV)
47. Deep Springs Valley (CA)
48. Desert NWR (NV)
49. Desert Tortoise NA (CA)
50. Don Edwards San Francisco Bay NWR (CA)
51. Eagle Lake (CA)
52. Elkhorn Slough NR (CA)
53. Encinitas (CA)
54. Farallon Is. (CA)
55. Forest of Nisene Marks SP (CA)
56. Fort Ord NMT (CA)
57. Giant Sequoia NMT (CA)
58. Golden Gate Bridge (CA)
59. Golden Trout Wilderness (CA)
60. Goose Lake (CA/OR)
61. Gray Lodge WLA (CA)
62. Great Basin NP (NV)
63. Great Valley Grasslands SP (CA)
64. Grizzly Island WLA (CA)
65. Grover Hot Springs SP (CA)
66. Half Moon Bay (CA)
67. Hayward Regional Shoreline (CA)
68. Henry W. Coe SP (CA)
69. Hickison Petroglyph RA (NV)
70. Hidden Cave (NV)
71. Hoopa Valley (CA)
72. Hwy. 1 (CA)
73. Hwy. 6 (NV)
74. Hwy. 50 (NV)
75. Hwy. 95 (NV)
76. Humboldt Lagoons SP (CA)
77. Humboldt Redwoods SP (CA)
78. Imperial NWR (AZ/CA)
79. Irish Beach (CA)
80. Ishi Wilderness (CA)
81. Jackpot (NV)
82. Joseph D. Grant CP (CA)
83. Joshua Tree NP (CA)
84. Kern NWR (CA)
85. Kettleman Hills (CA)
86. King Range NCA (CA)
87. Kings Canyon NP (CA)
88. Kruze Rhododendron SR (CA)
89. La Jolla (CA)
90. La Panza (CA)
91. Lake Henshaw (CA)

92. Lake Hughes (CA)
93. Lake Mead NRA (AZ/NV)
94. Lake Perris SRA (CA)
95. Lake Tahoe (CA/NV)
96. Lassen Volcanic NP (CA)
97. Lava Beds NMT (CA)
98. Little Panoche Rd. (CA)
99. Loma Prieta Mt. (CA)
100. Los Padres NF (CA)
101. Lovelock Valley (NV)
102. Lower Klamath NWR (CA)
103. Lost Lake (CA)
104. Lucia (CA)
105. Lytle Creek (CA)
106. Malibu (CA)
107. Marietta Wild Burro Range (NV)
108. Martinez (CA)
109. McCloud Falls (CA)
110. McCullough Range (NV)
111. Mecca Hills WA (CA)
112. Mendota WMA (CA)
113. Merced NWR (CA)
114. Modoc NF (CA)
115. Mojave NPR (CA)
116. Mono Lake (CA)
117. Monterey (CA)
118. Monterey Bay NMS (CA)
119. Monterey Submarine Canyon (CA)
120. Morro Bay SP (CA)
121. Moss Landing (CA)
122. Mt. Diablo SP (CA)
123. Mt. Shasta (CA)
124. Mt. Whitney (CA)
125. Mt. Whitney Fish Hatchery (CA)
126. Muir Woods NMT (CA)
127. Nacimiento-Fergusson Rd. (CA)
128. Newport Beach; Upper Newport Bay ER (CA)
129. Oakland (CA)
130. Onyx (CA)
131. Palo Alto Baylands PR (CA)
132. Picacho SRA (CA)
133. Pinnacles NP (CA)
134. Pit River (CA)
135. Piute Mts. (CA)
136. Pixley NWR (CA)
137. Point Arena (CA)
138. Point Lobos SNR (CA)

139. Point Pinos (CA)
140. Point Reyes NSS (CA)
141. Pyramid Lake (NV)
142. Redwood NP (CA)
143. Ruby Mts. (NV)
144. Sacramento NWR (CA)
145. Salinas River NWR (CA)
146. Salt Pt. SP (CA)
147. Salton Sea NWR (CA)
148. San Bernardino NF (CA)
149. San Clemente I. (CA)
150. San Gabriel Mts. NMT (CA)
151. San Jacinto Mts. (CA)
152. San Jacinto WLA (CA)
153. San Miguel I. (CA)
154. San Pablo Bay NWR (CA)
155. San Simeon (CA)
156. Sand Mt. RA (NV)
157. Santa Cruz I. (CA)
158. Santa Cruz Mts. (CA)
159. Santa Lucia Mts. (CA)
160. Santa Monica Mts. NRA (CA)
161. Santa Rosa (CA)
162. Santa Rosa & San Jacinto Mts. NMT (CA)
163. Santa Rosa I. (CA)
164. Searchlight (NV)
165. Sequoya NP (CA)
166. Sheldon NWR (NV)
167. Spring Mts. (NV)
168. Stebbins Cold Canyon Reserve (CA)
169. Stillwater NWR (NV)
170. Tecopa Hot Springs (CA)
171. Tehachapi Pass (CA)
172. Tioga Pass (CA)
173. Torrey Pines SR (CA)
174. Trinity Alps (CA)
175. Tule Elk SNR (CA)
176. Valley of Fire SP (NV)
177. Vic Fazio Yolo WLA (CA)
178. Vidal (CA)
179. Wellington Deer Range (NV)
180. Whiskeytown-Shasta-Trinity NRA (CA)
181. White Mts. (CA/NV)
182. Wildhorse Crossing Campground (NV)
183. Winnemucca (NV)
184. Yosemite NP

of the most impressive ecological communities on Earth. Three of the world's four tallest tree species grow here. California also has some surprisingly remote wilderness; a small population of Wolverines has managed to survive here virtually unnoticed for almost a century. Nevada, on the other hand, is sparsely populated and relatively little affected by human activities; only its southernmost part is seriously damaged by development.

Winters are generally cold and wet (drier in interior deserts); summers are hot and dry, but nights tend to be cool through most of the year. The Sierra Nevada typically gets massive amounts of snow in winter, so all roads across the mountains south of Lake Tahoe are closed from late fall until late spring. Some roads in the northern Coast Ranges are also subject to seasonal closures. It's a good idea to carry wheel chains in your car from October through late May. A cold current runs north-to-south along the coast, blanketing it with dense fog. The offshore waters are among the world's most productive, especially off Central California in summer months.

California and Nevada have approximately 200 mammalian species (the highest total in North America), 33 of them marine. Of these, 18 species (Mount Lyell Shrew, four chipmunks, Mohave Ground Squirrel, Nelson's Antelope Squirrel, three pocket mice, six kangaroo rats, Salt-marsh Harvest Mouse, and Sonoma Tree Vole) are endemic to California, one species (Palmer's Chipmunk) to Nevada, and six species (Inyo Shrew, two chipmunks, Mountain Pocket Gopher, Panamint Kangaroo Rat, and Pale Kangaroo Mouse) to both states. In addition, 7 or 8 species occur only in California and Mexico. Some widespread species, such as Bobcat and Gray Fox, are easier to see here than elsewhere.

The coast of northern California was once blanketed with old-growth redwood forests. Now only small fragments of them remain. Upslope and inland are mixed coniferous forests that gradually give way to open pinewoods and sagebrush flats. On the central coast, redwoods grow only in protected valleys opening to the sea; the mountains are covered with pine and oak forests and dense chaparral. Chaparral is supposed to burn once every few years, but nowadays these fires are suppressed at great cost to protect housing developments, so they happen less often and are more violent.

Redwood National Park and adjacent state parks form an outstanding natural area in far northwestern California. In addition to the world's tallest trees, there are numerous interesting mammals. Most of them are easier to find in mixed coniferous forests and coastal alder groves than in old-growth redwoods, although Coast Mole seems to prefer the latter. Look for Trowbridge's Shrew, Townsend's and Coast Moles, Fringed Myotis, American Black Bear, Black-tailed Deer, Roosevelt Elk (ask at the visitor center for herds' locations), Douglas's race of Pine Squirrel, Pacific Jumping Mouse, White-footed Vole, and Sonoma Tree Vole. The mountains immediately inland from the parks can be as wild as it gets; I've seen a Black Bear chasing cows there, and once had a Black Bear cub chase my car (I still don't know why).

If you follow the chain of "redwood parks" south, the next very scenic one is **Humboldt Redwoods State Park**, where Yellow-cheeked Chipmunk is easy to see and Sonoma Tree Vole is relatively common. The last park before San Francisco is **Muir Woods National Monument**, where Fog Shrew, Sonoma Chipmunk, and Pacific Jumping Mouse occur (most small mammals are more common at the periphery of the reserve, and even the most common ones can be difficult to find during weekends). If you have time, it's better to go to **Armstrong Redwoods State Natural Reserve**, which is open at night, has more diverse habitats, and is somewhat less overcrowded; campgrounds there are visited by semi-tame Striped Skunk and Gray Fox, while Yellow-cheeked Chipmunk occurs in shady forests. Coastal stretches of the very scenic **Highway 1** are good for night drives; look for Gray Fox, Bobcat, Northern Raccoon, Striped Skunk, and Virginia Opossum.

A bit farther inland, the rugged and remote **Trinity Alps** are a pristine mountain range where Pacific Shrew, Cougar, California Kangaroo Rat, and Western Red-backed Vole occur, although none is easy to find. More common mammals include Gray Fox, forest-adapted Black-tailed Jackrabbit, and Allen's Chipmunk.

Point Reyes National Seashore is another outstandingly scenic area on the coast of northern California. It is probably the easiest place in North America to see Bobcat and Gray Fox; look also for Long-tailed Weasel, Sonoma Chipmunk, and Pacific Jumping Mouse, as well as introduced Tule Elk and Fallow Deer. The summit plateau has the southernmost population of Sewellel. There are a few small Northern Elephant Seal and Harbor Seal rookeries in protected coves; the lighthouse is the best place to see Gray Whale from shore in winter (bring warm clothes at any time of year). The vicinity of the main visitor center is good for spotlighting.

Despite being densely populated with humans, the San Francisco

One of the best places to watch Northern Elephant Seals is at Highway 1 pullouts near San Simeon, California.

Bay Area has some interesting mammals. Harbor Porpoises can often be seen under the northern end of **Golden Gate Bridge**; California Sea Lions haul out at Pier 39 in downtown **San Francisco**, and city parks have Botta's Pocket Gopher, Mule Deer, Western Gray Squirrel, and California Ground Squirrel. Northern Raccoon and Striped Skunk frequent city streets at night, particularly in more "green" towns such as **Berkeley** and **Santa Rosa**.

One of the most interesting natural features of San Francisco Bay is winter high tides, which occur one to three times a year, each time over a period of two to three days. Tide charts can be found at saltwater tides.com/dynamic.dir/californiasites.html. When the tide floods coastal salt marshes, their unique inhabitants, including the tiny Salt-marsh Harvest Mouse and Ornate Shrew, can often be seen as they climb small shrubs or pieces of flotsam to escape the water. It's better if the highest tide is at dawn or dusk, but this doesn't happen every year. The most popular place to watch this is **Palo Alto Baylands Nature Preserve** (expect to meet lots of bird watchers looking for black and clapper rails). A less-known place is Lower Tubbs Island Trail in **San Pablo Bay National Wildlife Refuge**, where a different subspecies of Salt-marsh Harvest Mouse and Suisun Shrew (a beautiful silver-black race of Ornate Shrew) can be seen. The best habitat there is the marsh inside the loop part of the trail; it usually floods an hour or two later than the peak tide time listed in the charts. California Vole is abundant along the trail, particularly on levee tops, and nests of Western Harvest Mouse can be found in tall grass on levee slopes; some of these nests can be occupied by Vagrant Shrew. **Grizzly Island Wildlife Area** also has Salt-marsh Harvest Mouse and Suisun Shrew, plus Northern River Otter, Tule Elk, and Common Muskrat. Please note that the highest tide is a very stressful and dangerous time for mice and shrews (some of which are listed as endangered at the federal or state level), so it's better to keep to the trail and watch them from afar, preferably through a birding scope.

As you follow the coast south, stop at **Moss Landing** to look for Sea Otter in sloughs and for Common Bottlenose Dolphin offshore, and at **Año Nuevo State Park** to see abundant California Vole, Brush Rabbit, and (best in winter) Northern Elephant Seal. The **Santa Cruz Mountains** between Santa Cruz and Silicon Valley have an excellent network of narrow forest roads through redwood groves, oak forests, and dry chaparral. Look for Trowbridge's Shrew, American Shrew Mole, Broad-footed Mole, Bobcat, Cougar, Brush Rabbit, Merriam's Chipmunk, California Pocket Mouse, Narrow-faced Kangaroo Rat, and Western Harvest Mouse, as well as California, Piñon, and Brush Mice, and Dusky-footed Woodrat. **Forest of Nisene Marks State Park** is a particularly good location.

Pinnacles National Park is a nice hiking park where Townsend's Big-eared Bat and sometimes other bat species (including Long-eared Myotis and Western Bonneted Bat) roost in boulder caves. The very rare big-eared race of Narrow-faced Kangaroo Rat occurs in unburned areas of chamise chaparral with lots of bare ground, particularly near the eastern entrance and above Bear Gulch Reservoir. The park also has Bobcat, American Badger, Merriam's Chipmunk, Heermann's Kangaroo Rat (along Chalone Creek), and many other typical mammals of Central California. California Pocket Mouse and California, Deer, and Piñon Mice are common.

Monterey is probably the best city in the world for marine mammal watching. There is a California Sea Lion rookery near Fisherman's Wharf, a Harbor Seal rookery at the southern side of Hopkins Marine Station, and lots of Sea Otters just offshore. **Point Lobos State Natural Reserve** just south of Monterey is another good place to see Harbor Seal, California Sea Lion, and Sea Otter. Whales sometimes

California Sea Lion giving birth at Pier 39 in San Francisco.

enter Monterey Harbor, but it's better to join one or more whale-watching trips with **Monterey Bay Whale Watch**, which runs tours to **Monterey Bay National Marine Sanctuary**, particularly to the waters above the deep **Monterey Submarine Canyon**. In winter and spring the sea is often choppy and you mostly see Gray Whale, although Fin, Northern Minke, Baird's Beaked, and Killer Whales and a few dolphin species are also possible. The best time is late summer and early fall, when virtually all marine mammal species of the Pacific Coast can be encountered. Humpback, Blue, and Killer Whales;

In recent years, friendly Humpbacks have been recorded more and more often in Monterey Bay. They approach boats and play all kinds of tricks on whale watchers.

Pacific White-sided, Short-beaked Common, Risso's, and Northern Right Whale Dolphins; and Dall's and Harbor Porpoises are seen almost every week during that time. A few times a year, Monterey Bay Whale Watch runs two-day trips far offshore, where Sperm Whale and many species of beaked whales are also possible.

Other whale-watching opportunities in Central California include trips to the **Farallon Islands** (run by **San Francisco Whale Tours**), where Steller's Sea Lion can be seen from a distance, and to seamounts and submarine canyons farther north (run by **Shearwater Journeys** from **Bodega Bay**), where Cuvier's, Hubbs', Ginkgo-toothed, Perrin's, and Stejneger's Beaked Whales have been recorded. **Cordell Bank National Marine Sanctuary** and **Bodega Canyon** (a submarine canyon off Bodega Bay) seem to be particularly good locations for beaked whales, but it might be an artifact of their popularity among birders, who often go there to look for rare petrels and albatrosses.

Highway 1 south of Monterey is one of the world's most scenic roads (but avoid weekends and foggy days). At night you can see California Pocket Mouse and sometimes Cougar; look also for Sea Otter below scenic overlooks, Gray Whale offshore (in winter and spring), and Northern Elephant Seal at numerous haulouts north of San Simeon (watch for huge males crossing the highway!). A very steep, narrow, but paved **Nacimiento-Fergusson Road** branches off near Lucia and climbs the Santa Lucia Mountains. Once you get all the way up, turn left and go to the **Cone Peak Trail** trailhead. This is the most scenic short hike near the coast, and a good place to look for Merriam's Chipmunk, California Pocket Mouse, and California Mouse. Large-eared Woodrats occur at the trailhead, while tame Intermediate Woodrats emerge from the abandoned building at the summit at dusk. Narrow-faced Kangaroo Rat and Brush Mouse can be seen along the access road. Farther east, Nacimiento-Fergusson Road crosses an area where Cougar and Bobcat sightings are common.

Farther south, the **Channel Islands** lie offshore. Some are off-limits, others can be visited by boat tours from Ventura, while Catalina Island is accessible by ferries from Los Angeles. Of the accessible ones, Santa Cruz, San Miguel, and Catalina Islands have tiny Island Fox (a different subspecies on each island), which can often be seen around campgrounds, as well as distinctive races of Deer Mouse (the latter also occurs on Anacapa Island). Santa Cruz Island also has Ornate Shrew, Townsend's Big-eared Bat, Western Spotted Skunk, and Western Harvest Mouse (the latter three strictly nocturnal). Catalina Island has island races of Ornate Shrew, California Ground Squirrel, and Western Harvest Mouse. But the most interesting island is San Miguel, which has Western Spotted Skunk, huge numbers of California Sea Lion and Northern Elephant Seal, and a unique rookery at Point Bennett where California Sea Lion, Steller's Sea Lion, Northern Fur Seal, and sometimes Guadalupe Fur Seal haul out together. The island can be visited only with a ranger-guided tour, and you have to be in good physical shape to camp there and hike to the rookery (15 miles round trip); look up the details at the **Channel Islands National Park** website. Whales and dolphins are often encountered on these trips; Short-beaked Common Dolphin can often be seen in the thousands in summer months.

A less scenic but still interesting route from northern to southern California is through the Central Valley. Once a sea of grasslands, reedbeds, and riparian forests, the Central Valley has been almost entirely converted to agriculture, but small side valleys and oak savannas in surrounding foothills are better preserved. **Caswell Memorial State Park**, a remnant patch of floodplain woodlands, has rare riparian subspecies of Brush Rabbit and Dusky-footed Woodrat, as well as abundant Hoary Bat and Northern Raccoon. **Carrizo Plain National Monument** is an outstanding mammal-watching area with lots of unique species, particularly in the open grasslands of the southern part. Look for American Badger, Long-tailed Weasel (of the rare, very beautiful golden race), and Coyote at any time of the day; for Pronghorn, Tule Elk, California Ground Squirrel, and Nelson's Antelope Squirrel during daylight hours; for Canyon Bat, Pallid Bat, and Desert Cottontail at dusk; and on moonless nights for San Joaquin Kit Fox, Bobcat, Black-tailed Jackrabbit, Botta's

Brush Rabbit.

Pocket Gopher, San Joaquin Pocket Mouse, Southern Grasshopper Mouse, Desert Woodrat, and Giant, Heermann's, and Fresno Kangaroo Rats. The northern part of the park is also worth checking out, especially during the February-through-April wildflower season; look for Ornate Shrew and Western Harvest Mouse around vernal pools. Juniper- and oak-covered slopes around the valley have Cougar, Mule Deer, Merriam's Chipmunk, Deer Mouse, California Mouse, and Dusky-footed Woodrat. **Little Panoche Road** in San Benito County has many of these species. San Joaquin Kit Fox, San Joaquin Pocket Mouse, and all three kangaroo rats can be easier to find here in dry years. The best place for kangaroo rats is the gravel road branching off to the south, signposted "Panoche Hills BLM lands." **Tule Elk State Natural Reserve** has Tule Elk herds and is a good place to look for Long-tailed Weasel.

You can get from the Central Valley to the Mojave Desert by crossing **Tehachapi Pass**, where you can look for very rare White-eared Pocket Mouse, as well as for Great Basin Pocket Mouse (at lower elevations) and Panamint Kangaroo Rat. The southeastern side of the pass is better for all three. Once in the Mojave Desert, try **Desert Tortoise Natural Area** for endemic Mohave Ground Squirrel (spring only).

Lodgepole Chipmunk is endemic to the Sierra Nevada.

Deserts and mountains of Southern California are particularly rich in endemic species. **Joshua Tree National Park** has lots of typical desert mammals, such as Crawford's Desert Shrew, Coyote, Kit Fox, Bobcat, Desert Cottontail, Black-tailed Jackrabbit, White-tailed Antelope Squirrel, Round-tailed Ground Squirrel, Botta's Pocket Gopher, Merriam's and Desert Kangaroo Rats, Canyon and Cactus Mice, Southern Grasshopper Mouse, Desert Woodrat, and Long-tailed, San Diego, Spiny, and Desert Pocket Mice. California Chipmunk occurs in piñon-juniper forests; try the Pine City area. An isolated population of Chisel-toothed Kangaroo Rat inhabits high-elevation valleys, while Western Harvest Mouse occurs at Keys View. Palm-lined canyons near the southern entrance are particularly good for Canyon Bat and Western Yellow Bat.

Anza-Borrego Desert State Park is one of the most biologically diverse state parks in the U.S. A pond at the mouth of Borrego Palm Canyon attracts Canyon, Yellow, and Townsend's Big-eared Bats at dusk, and Desert Bighorn Sheep in late morning. Baja and Spiny Pocket Mice and Merriam's Kangaroo Rat occur in the area. To avoid summer heat, try **Cuyamaca Rancho State Park**, where Cougar is very common and both California and Merriam's Chipmunks occur. If you are serious about seeing all South California mammals, search **Mecca Hills Wilderness Area** for Western Yellow Bat, Western Small-footed Myotis, and possibly Spotted Bat, as well as Dulzura Kangaroo Rat and Desert Pocket Mouse; the hills north of **Lake Hughes** for Agile Kangaroo Rat, White-eared Pocket Mouse, and Northern Baja Mouse; and **San Jacinto Wildlife Area** for Stephens's Kangaroo Rat and San Diego Pocket Mouse.

The Sierra Nevada is popular with tourists throughout the year, and it's best to avoid places like Yosemite Valley during weekends and holidays, but the backcountry is a perfect place to get lost in the most scenic wilderness imaginable.

Lassen Volcanic National Park at the northern tip of the Sierra is the last place where Sierra subspecies of Red Fox is not vanishingly rare. It is also a good place to look for Preble's Shrew, American Marten, Ermine, American Pika, Least Chipmunk, Golden-mantled Ground Squirrel, Yellow-bellied Marmot, and Western Red-backed Vole.

Yosemite National Park has the world's highest chipmunk diversity; six species of chipmunks occur here. In mixed forests at lower elevations look for Canyon Bat, Coyote, and Western Gray Squirrel, Merriam's and Long-eared Chipmunks. Higher up, groves of giant sequoia, red fir, and other tall conifers are inhabited by Long-legged Myotis, American Marten, Northern Flying Squirrel, Douglas's race of Pine Squirrel, Western Jumping Mouse, and by Long-eared, Yellow-pine, Lodgepole, and Allen's Chipmunks. Alpine meadows along the road to Tioga Pass are inhabited by Mount Lyell Shrew, Yellow-bellied Marmot, Mountain Pocket Gopher, and Long-tailed Vole, while Belding's Ground Squirrel is abundant around the park entry gate at the pass. Sparsely vegetated granite slopes and mountaintops in the highest parts of the park support Alpine Chipmunk, Golden-mantled Ground Squirrel, and Western Heather Vole. Sierra Nevada Bighorn Sheep occur around Mono Pass and Rae Lakes; Alpine and Panamint Chipmunks and American Pika occur on a talus slope three miles up Mono Pass Trail. Ornate Shrew and Black Bear are common in Yosemite Valley. If Tioga Pass is closed, you can find some of those species at **Sonora Pass** to the north, or, if it is also closed, around **Lake Tahoe**. The southwestern side of the lake has extensive meadows and old-growth forests, where Ermine, Snowshoe

Hare, Allen's Chipmunk, Mountain Pocket Gopher, and American Beaver are also common. Try the vicinity of Taylor Creek Visitor Center for many of those, plus Least Chipmunk and (in October) American Black Bear.

Farther south, **Sequoia** and **Kings Canyon National Parks** have similar fauna. California Myotis is very common here in hollow trees and logs. Large-eared and Bushy-tailed Woodrats occur along rivers, while American Black Bear is easy to see in early summer in forest meadows around the Giant Forest Museum. An excellent side trip is to the part of Sequoia National Park called Mineral King; in summer it has American Black Bear, Alpine Chipmunk, and Yellow-bellied Marmot, while night drives along the access road are good for Trowbridge's Shrew, Bobcat, Cougar, American Badger, Western Spotted Skunk, and California, Brush, and Piñon Mice. Fisher is more common even farther south, for example, in the **Golden Trout Wilderness**.

Beyond Tioga Pass, a huge sagebrush desert called the Great Basin stretches all the way to Utah. It has rich and distinctive mammalian fauna. The southern shore of **Mono Lake** is good for Merriam's Shrew, Pygmy Rabbit, Desert Cottontail, Black-tailed Jackrabbit, Paiute Ground Squirrel, Dark Kangaroo Mouse, Great Basin and Little Pocket Mice, Merriam's, Chisel-toothed, and Panamint Kangaroo Rats, and Sagebrush Vole. A marsh on the northwestern side of the lake has unique desert populations of Broad-footed Mole and Sewellel, as well as Belding's Ground Squirrel, Mountain Cottontail, and (around the end of the boardwalk) Montane Vole. In **Ancient Bristlecone Pine Forest** in the White Mountains, Inyo Shrew, Panamint Chipmunk, and Golden-mantled Ground Squirrel occur among the world's oldest trees.

A spectacular region of deserts and mountains farther south along the California/Nevada state line is protected in **Death Valley National Park**. Parts of it are below sea level, so travel can be a bit challenging in summer heat, especially away from popular tourist routes. Look for Desert Kangaroo Rat in sand dunes, Panamint Kangaroo Rat and Desert Woodrat along the road to Dante's View, Panamint Chipmunk along Telescope Peak Trail, Canyon Mouse in narrow canyons, and Coyote everywhere. Abandoned mines inside the park have mostly been gated, but there are still plenty of ungated ones outside the park, and they have California and Fringed Myotis, Townsend's and Allen's Big-eared Bats, and Mexican Free-tailed Bat.

Deep Springs Valley, a little-known circular depression north of the park, has Pale Kangaroo Mouse (in sandy areas), White-tailed Jackrabbit, and Panamint Kangaroo Rat.

Nevada, a sea of sagebrush crossed by numerous mountain ranges, is a great country for night drives. There are plenty of narrow roads with almost no traffic. Crossing the state along **Highway 50** (signposted "America's loneliest road") or **Highway 6** (which has even less traffic late at night) is a good way to see Kit Fox, Pronghorn, Pygmy Rabbit, Dark and (rarely, in sandy areas) Pale Kangaroo Mice, and Ord's and Chisel-toothed Kangaroo Rats. Long-tailed Weasel and Rocky Mountain Elk are also seen occasionally. **Great Basin National Park**, an isolated, seldom-visited mountain range in eastern Nevada, has Inyo Shrew, American Pika, and Least and Uinta Chipmunks. Look also for Piute Ground Squirrel, Dark Kangaroo Mouse, and Chisel-toothed Kangaroo Rat along the access road.

Southern Nevada is covered with a different type of desert, dominated by creosote bush. White-tailed Antelope Squirrel and Black-tailed Jackrabbit are very common there. **Valley of Fire State Park** is the closest place to Las Vegas to see the colorful canyons typical for Utah. It is particularly good for White-tailed Antelope Squirrel; look also for Desert Bighorn Sheep just outside the main entrance and for Canyon Mouse in narrow gulches. The **Spring Mountains** are the only place where Palmer's Chipmunk lives (look at approximately 8,300 ft. [2,500 m] elevation, particularly around campgrounds, lodges, and ski resorts).

Northwestern U.S.

The five northwestern states are beautiful, diverse, and relatively well preserved. The coast has a mild but very wet climate; sunny breaks are most likely in late spring and early fall. The dry interior has moderately hot summers and cold winters. Some roads are closed in winter by heavy snowfalls, but high-elevation habitats can still be accessed by roads leading to ski resorts and to the stunning Crater Lake in Oregon.

More than 180 mammalian species (26 of them marine) inhabit the region. Of these, 10 (two shrews, three ground squirrels, two pocket gophers, two voles, and Olympic Marmot) are endemics. A few more species are shared only with northern California and/or extreme southwestern British Columbia. Coastal endemics such as

Coastal Northwestern U.S.

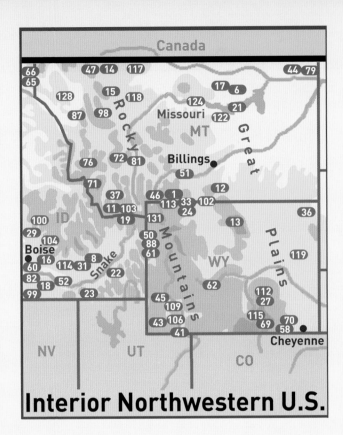

Interior Northwestern U.S.

Northwest

1. Absaroka-Beartooth Wilderness (MT)
2. Albany (OR)
3. Alfred A. Loeb SP (OR)
4. Alkali Lake (OR)
5. Altamont (OR)
6. American Prairie Reserve (MT)
7. Ankeny NWR (OR)
8. Atomic City (ID)
9. Bandon (OR)
10. Bend (OR)
11. Big Sheep Creek (MT)
12. Bighorn Canyon NRA (MT)
13. Bighorn Mts. (WY)
14. Blackfeet Tribe lands (MT)
15. Blackleaf WMA (MT)
16. Boise River WMA (ID)
17. Bowdoin NWR (MT)
18. Bruneau Dunes SP (ID)
19. Caribou-Targhee NF (ID)
20. Cascade-Siskiyou NMT (OR)
21. Charles M. Russell NWR (MT)
22. Cherry Springs NA (ID)
23. City of Rocks NR (ID)
24. Clarks Fork Canyon (WY)
25. Coquille River Falls Research NA (OR)
26. Columbia NWR (WA)

27. Como Bluff Dinosaur Graveyard (WY)
28. Conboy Lake NWR (WA)
29. Council (ID)
30. Crater Lake NP (OR)
31. Craters of the Moon NMT (ID)
32. Dean Creek Elk Viewing Area (OR)
33. Deep Lake (WY)
34. Deer Flat NWR (ID)
35. Devils Elbow SP (OR)
36. Devils Tower NMT (WY)
37. Dillon/Bannack SP (MT)
38. Emmett (ID)
39. Enterprise (OR)
40. Farragut SP (ID)
41. Flaming Gorge NRA (UT/WY)
42. Flume Creek Mountain Goat Viewing Area (WA)
43. Fort Bridger SHS (WY)
44. Fort Peck Reservation (MT)
45. Fossil Butte NMT (WY)
46. Gardiner (MT)
47. Glacier NP (MT)
48. Goose Lake SRA (CA/OR)
49. Grand Coulee (WA)
50. Grand Teton NP (WY)
51. Greycliff Prairie Dog Town SP (MT)
52. Hagerman Fossil Beds NMT (ID)
53. Hanford Reach NMT (WA)
54. Hart Mt. NWR (OR)
55. Hells Canyon NRA (ID/OR)
56. High Desert Museum (OR)
57. Humbug Mt. SP (OR)
58. Hutton Lake NWR (WY)
59. Illinois River Forks SP (OR)
60. Jacks Creek (ID)
61. Jackson (WY)
62. Jeffrey City (WY)
63. Julia Butler Hansen NWR (WA)
64. Kalmiopsis Wilderness (OR)
65. Kaniksu NF (ID)
66. Kootenai NWR (ID)
67. Lake Chelan NRA (WA)
68. Lake Coeur d'Alene (ID)
69. Lake Marie (WY)
70. Laramie River Greenbelt (WY)
71. Lee Metcalf NWR (MT)
72. Lewis and Clark Caverns SP (MT)
73. Lewis and Clark NHP (OR)
74. Little Pend Oreille NWR (WA)
75. Lolo Pass (OR)
76. Lost Creek SP (MT)
77. Lost Prairie Campground (OR)
78. Malheur NWR (OR)
79. Medicine Lake NWR (MT)

shrews, moles, and voles tend to be very difficult to see, but ground squirrels and the marmot are easy to find in the right season.

The coasts of Washington and Oregon are covered with forests of colossal conifers, mixed here and there with hardwood groves and wet meadows. These forests give way to more conventional coniferous forests farther inland and to lush alpine meadows at high elevations. The rainiest place is the Olympic Peninsula, famous for the moss-covered rainforests of **Olympic National Park**. Mammals are generally difficult to spot in such dense habitats, but look for Trowbridge's and Olympic Shrews, Keen's race of Long-eared Myotis, Roosevelt Elk, the never-changing race of Snowshoe Hare, Townsend's Chipmunk, Douglas's race of Pine Squirrel, Western Pocket Gopher, Pacific Jumping Mouse, and Keen's Mouse at low elevations, and for tame Black-tailed Deer, Mountain Goat (introduced), Yellow-pine Chipmunk, Golden-mantled Ground Squirrel, and endemic Olym-

80. Mima Mounds NA PR (WA)
81. Missouri Headwaters SP (MT)
82. Morley Nelson Snake River Birds of Prey NCA (ID)
83. Mt. Baker (WA)
84. Mt. Hood (OR)
85. Mt. St. Helens NMT (WA)
86. Mt. Rainier NP (WA)
87. National Bison Range (MT)
88. National Elk Refuge (WY)
89. Neptune SP (OR)
90. Nestucca Bay NWR (OR)
91. Newberry NMT (OR)
92. North Cascades NP (WA)
93. Oatman Flat (OR)
94. Olympia (WA)
95. Olympic NP (WA)
96. Oregon Dunes NRA (OR)
97. Oregon Caves NMT (OR)
98. Our Lake (MT)
99. Owyhee Uplands National Back Country Bwy. (ID/OR)
100. Payette NF (ID)
101. Prosser (WA)
102. Pryor Mt. Wild Horse Range (WY)
103. Red Rocks Lake NWR (MT)
104. Redfish Lake (ID)
105. Ridgefield NWR (WA)
106. Sage Junction (WY)
107. San Juan Islands NMT (WA)
108. Sea Lion Caves (OR)
109. Seedskadee NWR (WY)
110. Seep Lakes Unit of Columbia Basin WMA (WA)
111. Skagit WLA (WA)
112. Shirley Basin (WY)
113. Silver Gate (MT)
114. Silver Creek PR (ID)
115. Snowy Range Hwy. (Hwy. 130) (WY)
116. Steens Mt. (OR)
117. Sullivan Dam (MT)
118. Sun River Canyon (MT)
119. Thunder Basin NG (WY)
120. Toppenish NWR (WA)
121. Turnbull NWR (WA)
122. UL Bend NWR (MT)
123. Umatilla NWR (OR/WA)
124. Upper Missouri River Breaks NMT (MT)
125. Warner Wetlands (OR)
126. Weiser (ID)
127. Wenatchee (WA)
128. Wild Horse Island SP (MT)
129. Willapa NWR (WA)
130. Yakama Nation lands (WA)
131. Yellowstone NP (ID/MT/WY)

pic Marmot at Hurricane Ridge (accessible only in summer). Harbor Seal and (in winter and spring) Gray Whale are common along the western coast of the peninsula; look also for tame Western Spotted Skunks in Hole-in-the-Wall and Chilean Memorial campgrounds.

East across the Puget Sound, the spectacular **Mount Rainier National Park** has similar habitats but surprisingly different fauna. Look for Roosevelt Elk, Mountain Goat, American Pika, Cascade Ground Squirrel, Hoary Marmot, Water and Montane Voles, and Western Heather Vole above the timberline, and for Northern Flying Squirrel in tall conifers. On winter nights, beautiful black-and-silver Red Foxes visit ski lodges here.

Puget Sound is the most reliable place in North America to see Killer Whale, although there are way too many whale-watching boats per whale. Northern Minke and Gray Whales, Harbor Porpoise, and Steller's Sea Lion are also seen occasionally.

Roosevelt Elk during rut, Oregon.

Farther south, **Mount Saint Helens National Volcanic Monument** has mammals similar to those at Mount Rainier, but Creeping Vole at low elevations and Townsend's Chipmunk in the forests are easy to find. Little Brown Myotis, Townsend's Big-eared Bat, and Deer Mouse can occasionally be seen in lava tubes in the southern part of the park.

Eastern Washington State is much drier, and sagebrush dominates the intermontane valleys there. Mountain Goat and Desert Bighorn Sheep can be seen at **Flume Creek Mountain Goat Viewing Area** near Metaline.

The coast of Oregon is also very scenic. The most spectacular place is **Sea Lion Caves**, where dozens of Steller's Sea Lions and a few California Sea Lions haul out into a colossal sea cave. Baird's Shrew can occasionally be seen near cave entrances, and Gray Whales migrate offshore. Other good places along the coast include **Devils Elbow State Park**, where Trowbridge's, Fog, Pacific, Marsh, and Baird's Shrews as well as Townsend's Chipmunk occur; **Neptune State Park**, which has a Harbor Seal haulout; **Coquille River Falls Research Natural Area**, where Red Tree Vole is common; and **Alfred A. Loeb State Park**, where you have a chance of finding Northern Flying Squirrel, Dusky-footed Woodrat, Western Red-backed Vole, and Townsend's Vole.

A bit farther inland, **Columbia National Wildlife Refuge** has an endemic race of White-tailed Deer. In **Oregon Caves National Monument** near the California state line, Long-legged Myotis, Little Brown Myotis, and Townsend's Big-eared Bat can sometimes be seen during winter cave tours. Two other excellent places for mammal watching near the state line are **Cascade-Siskiyou National Monument** and the remote **Kalmiopsis Wilderness**. Both have Allen's and Siskiyou Chipmunks and California Kangaroo Rat.

Willamette Valley between the coastal ranges and the Cascade Mountains has two endemic mammal species: Camas Pocket Gopher and Gray-tailed Vole. Both are easy to see in agricultural areas, particularly after heavy rains; they are also present in **Ankeny National Wildlife Refuge**.

One of the most beautiful parks on the continent is **Crater Lake National Park** in the Cascades. In addition to stunning vistas, it has Western Gray Squirrel; Western Red-backed, Long-tailed, and Water Voles; Northern Porcupine; and numerous other mammals. On winter nights, American Marten can sometimes be seen near the lodge at the crater rim. Farther north, **Lolo Pass** at Mt. Hood is the best place to see Sewellel, the world's most ancient living rodent.

Sagebrush deserts of eastern Oregon have totally different wildlife. Two beautiful, remote, and pristine mountains rise from the sea

The amazing Sea Lion Caves, Oregon.

More than half of the remaining Pronghorn (more than one million) live in Wyoming.

of sagebrush: **Steens Mountain** and **Hart Mountain**, where Desert Bighorn Sheep, Pronghorn, Pygmy Rabbit, Merriam's and Wyoming Ground Squirrels, and Great Basin Pocket Mouse can be found.

The northern part of Great Basin in eastern Oregon and southern Idaho, plus similar grasslands of southeastern Washington, have the world's highest diversity of ground squirrels, although they are not all in the same place and a lot of driving is required to see them all. See species accounts in Part III for particular locations. One good place is **Snake River Birds of Prey National Conservation Area**, which has a few species of ground squirrels and lots of American Badgers. **Hells Canyon National Recreation Area**, also in Idaho, has Merriam's Shrew and Bushy-tailed Woodrat. **Owyhee Uplands Back Country Byway**, running between Grand View in Idaho and Jordan Valley in Oregon, offers a good chance to see Coyote, Badger, Bobcat, Pronghorn, Belding's and Merriam's Ground Squirrels, Least Chipmunk, and possibly also Long-tailed Weasel, Cougar, and Pygmy Rabbit. **Silver Creek Preserve** near Picabo (Idaho) is famous for spectacular population explosions of Montane Vole, occurring once every few years and attracting all kinds of predators. It also has

Long-tailed Weasel, Badger, Striped Skunk, Rocky Mountain Elk, Moose, Beaver, Columbian Ground Squirrel, Yellow-bellied Marmot, and Bushy-tailed Woodrat.

A particularly good place for mammal watching is **City of Rocks National Reserve** in southeastern Idaho. Its rock outcrops provide shelter for Big Brown and Spotted Bats; Townsend's Big-eared Bat; Little Brown, Long-eared, Western Small-footed, and Fringed Myotis; Western Spotted Skunk; Ermine; Cougar; Cliff and Least Chipmunks; Canyon Mouse; and Bushy-tailed Woodrat, while surrounding grasslands and creeks are inhabited by Merriam's, Vagrant, and Cordilleran Water Shrews; Long-tailed Weasel; Pygmy Rabbit; White-tailed Jackrabbit; Belding's and Richardson's Ground Squirrels; Yellow-bellied Marmot; Great Basin Pocket Mouse; Ord's Kangaroo Rat; Northern Pocket Gopher; Western Harvest Mouse; Northern Grasshopper Mouse; Long-tailed, Montane, and Sagebrush Voles; Western Jumping Mouse, and many other species.

The Rocky Mountains separate the Great Basin from the short-grass prairies of the Great Plains in eastern Montana and Wyoming. Situated on a volcanic plateau with a gigantic caldera (extra-large crater) in the middle, **Yellowstone National Park** (mostly in Wyoming) is probably the best mammal-watching site in the Lower 48 states. Accommodation inside the park is very limited, so it is better to stay just outside the northeastern entrance, close to Lamar Valley,

American Badger is most often seen in the vicinity of prairie dog towns and ground squirrel colonies, but it can also survive on mice and voles.

where Gray Wolf, Grizzly Bear, American Badger, Pronghorn, Uinta Ground Squirrel, and American Beaver are common. Note that Lamar Valley is open for hiking, although most visitors are too scared of large animals to use its excellent trails and prefer to endure traffic jams on the road. Rocky Mountain Bighorn Sheep are also seen there sometimes, although the Tower Junction area is better for them. American Black Bear, Moose, and Mule Deer are often visible from the northeast entrance road. Mountain Goat can be seen with good binoculars or a birding scope from Pebble Creek, Barronette Peak, and Trout Lake parking lots. Coyote, Rocky Mountain Elk, Plains Bison, and Northern Porcupine are common throughout most of the park, although the largest herds of Bison are usually in Lamar Valley and the central area. Northern River Otter lives in Lake Yellowstone and in Trout Lake. Small ponds around Trout Lake also have a sizeable population of Water Vole. American Marten, Long-tailed Weasel, Ermine, American Pika, and Uinta Chipmunk occur along Mt. Washburn, Lost Lake, and Hellroaring Creek Trails. Other small mammals include Masked Shrew, American Water Shrew, Least and Yellow-pine Chipmunks, Pine Squirrel, Golden-mantled Ground Squirrel, Northern Pocket Gopher, Deer Mouse, and Bushy-tailed Woodrat, as well as Long-tailed, Meadow, Western Heather, and Southern Red-backed Voles.

Mammal watching in Yellowstone in winter is a very different experience. Much of the park can be visited only with an organized tour at that time, but the best road for viewing wildlife, the one between the northern and northeastern entrances, is open. As a matter of fact, it is open even during the few months in spring and fall when the rest of the park is completely closed. In winter, snow tires are recommended and sometimes required, but you can still drive the entire road in a passenger car in less than two hours, and, unlike in summer, there is very

Montane Vole is one of the most frequently seen rodents of western mountains.

little traffic. Plains Bison and Rocky Mountain Elk are visible almost everywhere; American Marten and Snowshoe Hare are common around Mammoth Hot Springs; Rocky Mountain Bighorn Sheep graze above the road between Mammoth and Tower Junction; Gray Wolf, Coyote, and American Badger are often seen in Lamar Valley; Moose and Mountain Goat can sometimes be spotted a few miles before the northeastern entrance. Wolverines living in high mountains along the northern and eastern borders of the park descend to lower elevations in winter. Another year-round access road is Highway 191, which runs along the western edge of the park; it crosses some excellent Western Heather Vole habitat about 20 miles north of West Yellowstone. A bit farther west, at the edge of Yellowstone Plateau, is **Red Rock Lakes National Wildlife Refuge**, where Moose, White-tailed Deer, White-tailed Jackrabbit, Wyoming Ground Squirrel, and Idaho Pocket Gopher are easy to see.

Unfortunately, Yellowstone National Park was created in such a way that it doesn't include enough low-elevation land to provide its inhabitants with winter pasture, so many of its larger animals have to move out in winter and descend into unprotected areas where hunters eagerly await. Pronghorn mostly migrate east, to Wyoming grasslands; Elk move south, while Bison move north into Montana, where ranchers accuse them of spreading bovine tuberculosis to cattle (even though such transmission has never been confirmed) and insist on culling them. Highway 89 from Livingston to the park's northern entrance is a good place to see all these species, as well as Mule Deer and Rocky Mountain Bighorn Sheep, in winter months.

South of Yellowstone, **Grand Teton National Park** and the **National Elk Refuge** (the main wintering area of Yellowstone Elk) have similar fauna, but some species such as Moose and Cougar are easier to find there, while American Marten is regularly seen along Cascade Canyon Trail and around Jenny Lake. A hill inside the National Elk Refuge usually has Rocky Mountain Bighorn Sheep in winter. Moose-Wilson Road is good for Common Muskrat and American Beaver, while Moose Ponds Trail has Mule Deer, Yellow-bellied Marmot, Uinta Ground Squirrel, and Yellow-pine Chipmunk in summer.

There is a useful online forum where recent sightings in and around Yellowstone are often posted (http://forums.yellowstone.net/viewforum.php?f=15).

Other interesting places in Wyoming include **Hutton Lake National Wildlife Refuge** (look for American Badger, Wyoming

Plains Bison are slowly coming back to the Great Plains, but there is a lot of work to do before the migrations of their large herds can be restored.

Ground Squirrel, and White-tailed Prairie Dog); **Como Bluff Dinosaur Graveyard** (a great place to see Wyoming and Spotted Ground Squirrels in the company of Black-tailed Prairie Dog); the area south of **Jeffrey City**, where endemic Wyoming Pocket Gopher occurs; and **Thunder Basin National Grassland**, which is very good for short-grass prairie species such as Merriam's Shrew, Pronghorn, Desert and Mountain Cottontails, White-tailed Jackrabbit, Merriam's and Thirteen-lined Ground Squirrels, Black-tailed Prairie Dog, Olive-backed Pocket Mouse, Northern Grasshopper Mouse, Deer Mouse, Meadow Vole, and Sagebrush Vole. Pronghorn are abundant almost everywhere in the eastern half of the state, and Mule Deer can be seen in huge herds in a few places in the southwestern part (see species account in Part III for a list of locations).

Montana is even more pristine than Wyoming. The most scenic place here is **Glacier National Park**. It is a very diverse park, with rainforest-type conifers on the western slopes, arid foothills on the eastern side, and extensive alpine areas in between. It has Grizzly and American Black Bears, Rocky Mountain Elk, Rocky Mountain Bighorn Sheep, Mountain Goat, Snowshoe Hare, Red-tailed Chipmunk (look around campgrounds on the western side and at Avalanche Lake), Richardson's and Golden-mantled Ground Squirrels, and

Southern Red-backed Vole (try Hole-in-the-Wall Trail), among other mammals. It is also the only place in the Lower 48 states where your chances of seeing Wolverine are not infinitely small. Hidden Lake Overlook Trail is good for Columbian Ground Squirrel, Long-tailed Weasel, Hoary Marmot, and American Pika. The food preparation area in Bowman Lake Campground can have Deer Mouse and Western Jumping Mouse at night. The breathtakingly beautiful Granite Park Chalet Trail is particularly good for seeing large mammals such as Grizzly Bear against the backdrop of snow-clad peaks and blue glaciers. Mountain Cottontail and White-tailed Deer are abundant along wood margins in lowland areas around the park. Much of the park is accessible only in July and August.

For Great Plains species, one of the best places in Montana is the land of the **Blackfeet Tribe**, where endless grasslands are inhabited by Prairie and Preble's Shrews, Swift Fox, and Pronghorn, while Northern Porcupine is common in aspen groves. **National Bison Range** has lots of Plains Bison, as well as White-tailed Deer and Water Vole along streams. Farther east, the large **Charles M. Russell National Wildlife Refuge** is the base for the future **American Prairie Reserve**, which is hoped to eventually restore an immense chunk of shortgrass prairie. This is a slow, painstaking, and expensive effort, but in a few decades you might be able to see large migrating herds of Plains Bison and Pronghorn once again. The Refuge and the Reserve have also Prairie Shrew, Cougar, Bobcat, Black-footed Ferret, Long-tailed Weasel, American Mink, American Badger, Swift Fox, Red Fox,

Prairie dog colonies are magnets for all kinds of grassland creatures, from tiny beetles to American Buffalo that wallow in the sand there.

Coyote, Rocky Mountain Bighorn Sheep, Rocky Mountain Elk, Mule and White-tailed Deer, and numerous rodents. As you drive or walk across the prairie, don't forget to stop searching for rare wildlife once in a while to simply look around and take in the beauty of the Plains. Endless rolling grasslands are becoming a rare sight on our planet nowadays.

Southwestern U.S.

Among the world's deserts, none comes anywhere close to the canyon country of the American Southwest in terms of natural beauty. Canyons, deserts, coniferous forests, cactus groves, and alpine meadows of the Southwest provide spectacular habitats for numerous mammals, from tropical species such as White-nosed Coati to northerners such as Moose. Although many of the best sites are easily accessible by car, you will enjoy this land a lot more if you do a few long hikes and horseback trips into the wilderness.

The Southwest is an arid country, with rains falling mostly in late winter/early spring and then again (but not every year and not everywhere) in late summer. At higher elevations, a lot of snow accumulates by the end of winter. Flash floods are a serious danger during thunderstorms or on warm spring days when the snow melts. Summer days can be extremely hot, and winter nights are usually freezing, except in a few places along the Mexico border where freezes are rare.

This region has the highest diversity of land mammals in North America, with at least 184 species. Of these, 9 (Abert's Squirrel, two prairie dogs, two chipmunks, two pocket gophers, Stephens's Woodrat, and Black-eared Mouse) are endemic, and 16 are shared only with Mexico.

Western Utah is a part of the Great Basin, very similar to the Nevada deserts. The best place to see sagebrush species such as Pygmy Rabbit and Great Basin Pocket Mouse is the remote **Fish Springs National Wildlife Refuge**. Uinta Ground Squirrel can be seen around **Utah Lake** on the eastern edge of the desert, while Long-tailed Pocket Mouse is common around the Great Salt Lake—for example, in **Antelope Island State Park**. That park also has Coyote, Bobcat, American Badger, Mule Deer, Plains Bison, Pronghorn, Rocky Mountain Bighorn Sheep, and Chisel-toothed Kangaroo Rat.

A mountain range with fauna typical of the Rockies separates the

White-nosed Coati, Cave Creek Canyon, Arizona.

Great Basin from the immense Colorado Plateau that covers eastern Utah, western Colorado, northern Arizona, and northwestern New Mexico. Thousands of colorful canyons crisscross the plateau; no two of them are alike. Some are very touristy; others are no less scenic, but little known, such as the outstanding slot canyons and slick rock valleys of **Grand Staircase-Escalante National Monument** and the endless canyon labyrinths of **Canyonlands National Park**. Only a few mammals are canyon specialists, but even the most common and widespread animals look particularly beautiful when encountered in these spectacular places. Crevices, niches, and small caves in canyon walls (particularly abundant in parts of **Capitol Reef National Park**) shelter California, Yuma, and Long-legged Myotis; Canyon, Spotted, Townsend's Big-eared, Pallid, Mexican Free-tailed, and Big Free-tailed Bats; Ringtail; Gray Fox; Hopi, Cliff, and Uinta Chipmunks; Rock Squirrel; Golden-mantled Ground Squirrel; and Bushy-tailed Woodrat. Canyon Mouse, Desert Bighorn Sheep, Mule Deer, Cougar, and Long-tailed Weasel occur around springs. Sandy washes and sagebrush flats are inhabited by White-tailed Antelope Squirrel, Plains Pocket Mouse, and Ord's Kangaroo Rat. **Bryce Canyon National Park** has also Pronghorn, Least Chipmunk, and Utah Prairie Dog (there is a large prairie dog town near the visitor center). Abandoned uranium mines around **Temple Mountain** (near the

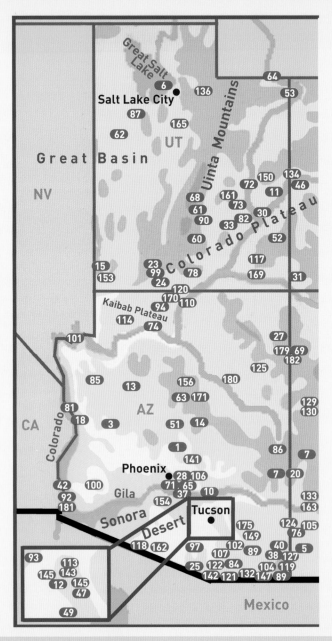

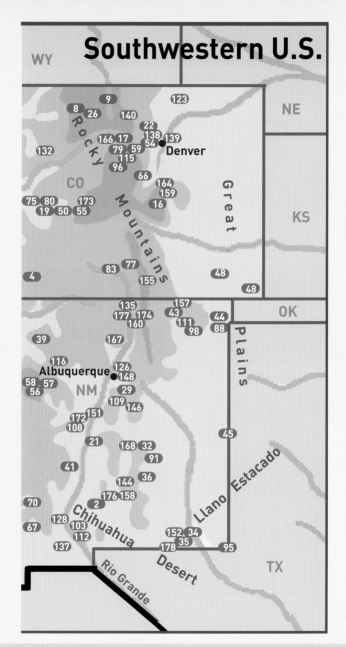

Southwestern U.S.

WY

NE

9
8 26
140
123
22
138 139
166 17
54
79 59
Denver
115
132
96
CO
66
164
159
75 80 173
16
19 50 55

Rocky Mountains

Great

KS

4
83 77
155
48
48

135
157
177 174
43
160
111
44
39
167
98
88

Plains

OK

116
126
Albuquerque 148
58 57
56
NM
29
109 146
172 151
108
21
168 32
91
41
144 36
176 158
70
2
128
67
103
112
137

Chihuahua

45

Llano Estacado

152 34
35
178
95

Desert

TX

Rio Grande

Southwest

1. Agua Fria NMT (AZ)
2. Aguirre Springs Campground (NM)
3. Alamo Lake SP (AZ)
4. Animas Mt. Trail (CO)
5. Animas Valley (NM)
6. Antelope Island SP (UT)
7. Apache Sitgreaves NFs (AZ/NM)
8. Arapaho NF (CO)
9. Arapaho NWR (CO)
10. Aravaipa Canyon (AZ)
11. Arches NP (UT)
12. Arizona-Sonora Desert Museum (AZ)
13. Aubrey Valley (AZ)
14. Baker Lake (AZ)
15. Beaver Dam Wash NCA (UT)
16. Beaver Creek Reservoir SWA (CO)
17. Bighorn Viewing Area (CO)
18. Bill Williams River NWR (AZ)
19. Black Canyon of the Gunnison NP (CO)
20. Blue Range WA (NM)
21. Bosque del Apache NWR (NM)
22. Boulder (CO)
23. Bryce Canyon NP (UT)
24. Buckskin Gulch (UT)
25. Buenos Aires NWR (AZ)
26. Cameron Pass (CO)
27. Canyon de Chelly NMT (AZ)
28. Canyon Lake (AZ)
29. Canyon Largo (NM)
30. Canyonlands NP (UT)
31. Canyons of the Ancients NMT (CO)
32. Capitan Mts. WA (NM)
33. Capitol Reef NP (UT)
34. Carlsbad (NM)
35. Carlsbad Caverns NP (NM)
36. Carr Gap Canyon (NM)
37. Casa Grande Ruins NMT (AZ)
38. Cave Creek Canyon (AZ)
39. Chaco Culture NHP (NM)
40. Chiricahua NMT (AZ)
41. Chloride (NM)
42. Cibola NWR (AZ)
43. Cimarron Canyon SP (NM)
44. Clayton Lake SP (NM)
45. Clovis (NM)
46. Colorado NMT (CO)
47. Colossal Cave (AZ)
48. Comanche NG (CO)
49. Continental (AZ)
50. Curecanti NRA (CO)
51. Dead Horse Ranch SP (AZ)
52. Devils Canyon Campground (UT)
53. Dinosaur NMT (CO/UT)
54. Dinosaur Ridge (CO)
55. Doyleville (CO)
56. El Malpais NCA (NM)
57. El Malpais NMT (NM)
58. El Morro NMT (NM)
59. Elk Meadow Park (CO)
60. Escalante Petrified Forest SP (UT)
61. Fish Lake (UT)
62. Fish Springs NWR (UT)
63. Flagstaff Arboretum (AZ)
64. Flaming Gorge NRA (UT/WY)
65. Florence Junction (AZ)
66. Florissant Fossil Beds NMT (CO)
67. Fort Bayard (NM)
68. Fremont Indian SP (UT)
69. Gallup (NM)
70. Gila Cliff Dwellings NMT (NM)
71. Gilbert (AZ)
72. Green River (UT)
73. Goblin Valley SP (UT)
74. Grand Canyon NP (AZ)
75. Grand Mesa (CO)
76. Granite Gap (NM)
77. Great Sand Dunes NP (CO)
78. Grand Staircase-Escalante NMT (UT)
79. Guanella Pass (CO)
80. Gunnison Gorge NCA (CO)
81. Havasu NWR (AZ)
82. Henry Mts. (UT)
83. Hot Creek WA (CO)
84. Huachuca Mts. (AZ)
85. Hualapai Mts. (AZ)
86. Hwy. 191 (AZ)
87. Hwy. 196 (UT)
88. Hwy. 56 (NM)

89. Hwy. 80 (AZ)
90. Hogan Pass (UT)
91. Hondo Valley (NM)
92. Imperial NWR (AZ/CA)
93. Ironwood Forest NMT (AZ)
94. Jacob Lake (AZ)
95. Jal (NM)
96. Kenosha Pass (CO)
97. Kitt Peak (AZ)
98. Kiowa NG (NM)
99. Kodachrome Basin SP (UT)
100. Kofa NWR (AZ)
101. Lake Mead NRA (AZ/NV)
102. Las Cienegas NCA (AZ)
103. Las Cruces (NM)
104. Leslie Canyon NWR (AZ)
105. Lordsburg Playas (NM)
106. Lost Dutchman SP (AZ)
107. Madera Canyon (AZ)
108. Magdalena Mts. (NM)
109. Manzano Mts. (NM)
110. Marble Canyon (AZ)
111. Maxwell NWR (NM)
112. Mesquite (NM)
113. Mt. Lemmon (AZ)
114. Mt. Trumbull Trail (AZ)
115. Mt. Evans (CO)
116. Mt. Taylor (NM)
117. Natural Bridges NMT (UT)
118. Organ Pipe Cactus NWR (AZ)
119. Paramore Crater (AZ)
120. Paria Canyon-Vermilion Cliffs Wilderness (AZ/UT)
121. Parker Canyon Lake (AZ)
122. Patagonia Lake SP (AZ)
123. Pawnee NG (CO)
124. Peloncillo Mts. (AZ/NM)
125. Petrified Forest NP (AZ)
126. Placitas (NM)
127. Portal (AZ)
128. Prehistoric Trackways NMT (NM)
129. Quemado (NM)
130. Quemado Lake (NM)
131. Rabbit Valley (CO)
132. Ramsey Canyon PR (AZ)
133. Redrock (NM)
134. Rifle Falls SP (CO)
135. Rio Grande del Norte NMT (NM)

136. Rockport SP (UT)
137. Rock Hound SP (NM)
138. Rocky Flats NWR (CO)
139. Rocky Mt. Arsenal NWR (CO)
140. Rocky Mt. NP (CO)
141. Roosevelt Lake WLA (AZ)
142. Ruby Rd. (AZ)
143. Sabino Canyon RA (AZ)
144. Sacramento Mts. (NM)
145. Saguaro NP (AZ)
146. Salinas Pueblo Missions NMT (NM)
147. San Bernardino NWR (AZ)
148. Sandia Crest (NM)
149. Santa Rita Mts. (AZ)
150. Sego Canyon (UT)
151. Sevilleta NWR (NM)
152. Sitting Bull Falls RA (NM)
153. Snow Canyon SP (UT)
154. Sonoran Desert NMT (AZ)
155. Spanish Peaks (CO)
156. Sunset Crater Volcanic NMT (AZ)
157. Sugarite Canyon SP (NM)
158. Sunspot (NM)
159. Tamarack Ranch SWA (CO)
160. Taos (NM)
161. Temple Mt. (UT)
162. Tohono O'Odham Nation (AZ)
163. Turkey Creek Rd. (NM)
164. U.S. Air Force Academy (CO)
165. Utah Lake SP (UT)
166. Vail (CO)
167. Valles Caldera NPR (NM)
168. Valley of Fires RA (NM)
169. Valley of the Gods (UT)
170. Vermilion Cliffs NMT (AZ)
171. Walnut Canyon NMT (AZ)
172. Water Canyon (NM)
173. Waunita Hot Springs (CO)
174. Wheeler Peak (NM)
175. Whetstone Mts. (AZ)
176. White Sands NMT (NM)
177. Wild Rivers RA (NM)
178. Wilderness Ridge (NM)
179. Window Rock (AZ)
180. Woods Canyon Lake (AZ)
181. Yuma (AZ)
182. Zuni Pueblo (NM)

Texas Antelope Squirrel is among the very few mammals capable of remaining active in the open in the midday heat of the Chihuahuan Desert.

wonderful **Goblin Valley State Park**) are inhabited by Big Brown and Pallid Bats. Another special site is the Wolfe Ranch in unbelievably beautiful **Arches National Park**, where numerous bats swarm above a small creek on summer evenings; most of them are Canyon Bats, but there are also Yuma and Western Small-footed Myotis and Pallid Bats. The campground area of **Escalante Petrified Forest State Park** is a good place to look for Western Red Bat.

In the Arizona part of the Colorado Plateau, the most scenic area lies just north of Marble Canyon; it includes **Paria Canyon** and **Vermilion Cliffs National Monument**, where Arizona Myotis, Spotted Bat, and Chisel-toothed Kangaroo Rat occur.

The **Grand Canyon** separates the Plateau from the rest of Arizona. Its North Rim is higher and much more densely forested, and is famous for the largest subspecies of Mule Deer and the most beautiful subspecies of Abert's Squirrel, the so-called Kaibab Squirrel with a gorgeous white tail. Coyote, Colorado Chipmunk, and Bushy-tailed Woodrat are also common there. Permanent ponds near Jacob Lake are visited by many bat species, including Allen's Big-eared and Spotted Bats. Cliff Chipmunk, Golden-mantled Ground Squirrel, Rock Squirrel, and Brush Mouse are common on both rims, while Spotted Ground Squirrel and a different race of Abert's Squirrel live on the South Rim (look for the ground squirrel around the airport). Inaccessible bat caves are located under the Mohave Point and Abyss overlooks on the South Rim; walking along the rim edge at dusk, you can see California Myotis, Pallid Bat, and Mexican Free-tailed and Big Free-tailed Bats flying around. Phantom Ranch at the bottom of the canyon has tame Ringtail and Western Spotted Skunk; there are also Spotted Bat roosts in surrounding rock walls, and Desert Big-horn Sheep are often seen on the slopes above.

Another outstanding place in northern Arizona is **Aubrey Valley**, an area of desert grasslands west of Seligman. Large towns of Gunnison's Prairie Dog attract numerous American Badgers and Coyotes. This is one of the most accessible Black-footed Ferret reintroduction sites; a night of spotlighting along Route 66 might result in seeing this exceptionally rare predator. Pronghorn, Collared Peccary, Eastern and Desert Cottontails, Bushy-tailed Woodrat, Northern and Southern Grasshopper Mice, and Ord's Kangaroo Rat also occur in the valley. The area between mile markers 130 and 120 on Route 66 is the best for night drives and spotlighting.

Mule Deer got its name for its large, mulelike ears. Deer in the Southwest are particularly big-eared, while those in the Northwest have more conventional ear size.

The **Arboretum at Flagstaff**, accessible by taking Route 66 east from exit 191 on I-40 and then turning south onto Woody Mountain Road, is an excellent site to see Gray-collared Chipmunk. Rocky Mountain Elk, Abert's Squirrel, and Golden-mantled Ground Squirrel are also common in that area. **Petrified Forest National Park** has more Pronghorn (sometimes very tame), and various buildings there are often used by roosting bats; ask park rangers for the best places to look for California and Yuma Myotis and Pallid Bat. There is a colony of Gunnison's Prairie Dog at Newspaper Rock turnoff. Cave Myotis occurs in ancient ruins in **Walnut Canyon National Monument**, where Abert's Squirrel and Gray-collared Chipmunk are also common.

For a mammal watcher, southern Arizona and extreme southwestern New Mexico are among the most interesting parts of the continent. This is the northern section of the Sonoran Desert, which continues into Mexico and is inhabited by lots of endemic species. Groves of giant cacti and warm canyons with pine-oak forests have wonderfully diverse, mostly tropical mammal, bird, and reptile fauna. The so-called "sky islands" (isolated mountain ranges) are inhabited by animals of northern origin.

The city of **Yuma** is located in the hottest corner of North America. Desert Pocket Mouse is common around town. Spiny Pocket Mouse and Hispid Cotton Rat occur in nearby **Cibola National Wildlife Refuge**. Desert Bighorn Sheep, Southern Grasshopper Mouse, and Arizona Woodrat can be found in **Kofa National Wildlife Refuge**. Look also for beautifully colored Yuma Cougar and particularly big-eared Black-tailed Jackrabbit in this area.

Phoenix has a nice zoo and botanical garden where wild Pallid Bat, Desert Cottontail, Black-tailed Jackrabbit, Round-tailed Ground Squirrel, Harris's Antelope Squirrel, and Arizona Cotton Rat occur on the grounds. In summer it is better to start your visit from nearby Papago Park because, unlike the zoo and the botanical garden, it opens at dawn. You can also watch the evening emergence of thousands of Mexican Free-tailed Bats from cracks in the walls of the University of Phoenix Stadium.

Tucson is surrounded by excellent habitat on all sides. **Saguaro National Park** is a great place to look for Mexican Long-tongued Bat (around hummingbird feeders and flowering cacti), Round-tailed Ground Squirrel, Harris's Antelope Squirrel, Merriam's Kangaroo Rat, and Bailey's Pocket Mouse. **Colossal Cave** has Lesser Long-

Rocky Mountain Bighorn Sheep is the easiest wild sheep in North America to see. Colorado has a few places where it can be found most of the time.

nosed Bat and Southwestern Myotis, and its entrance area is frequented by Mexican Woodrat. **Sweetwater Wetlands**, a sewage treatment plant just off I-10 in Tucson, is a good place to look for Arizona Cotton Rat and Bobcat.

Organ Pipe Cactus National Monument has the most diverse cactus forests north of the Mexico border, and some of the best night driving. Look for Crawford's Desert Shrew, Underwood's Bonneted Bat, California Leaf-nosed Bat, Collared Peccary, Antelope Jackrabbit, Harris's Antelope Squirrel, Round-tailed Ground Squirrel, Arizona and Silky Pocket Mice, Merriam's Kangaroo Rat, Cactus Mouse, and White-throated Woodrat. Reptiles are also numerous. Quitobaquito Springs is a particularly good watering hole where many bats and other species can be observed. Currently this area can be accessed only with sporadic daytime ranger-led tours, but it might become open again if border-related concern subsides.

Desert grasslands and woodlands of **Buenos Aires National Wildlife Refuge** have Bobcat, Pronghorn of the rare Sonoran subspecies, Animas race of Botta's Pocket Gopher, and Mesquite Mouse.

In **Patagonia Lake State Park**, Arizona Cotton Rat and Western Harvest Mouse are often seen around the bird feeders at the visitor center.

Four canyons in the southeastern corner of Arizona are famous among bird watchers, and have some great mammals as well: **Madera Canyon** in the Santa Rita Mountains, **Ramsey Canyon** in the Huachuca Mountains, and **Leslie** and **Cave Creek Canyons** in the Chiricahua Mountains. All except Leslie have lodges where hummingbird feeders are often visited by Mexican Long-tongued and Lesser Long-nosed Bats. Madera Canyon has Arizona Shrew, Hooded Skunk, and Ringtail. Bird feeders at Santa Rita Lodge there are often visited by Arizona Gray Squirrel and Rock Squirrel. Look also for Antelope Jackrabbit, the Animas race of Botta's Pocket Gopher, and Bailey's Pocket Mouse along the access road. Ramsey Canyon has Arizona Gray Squirrel and Collared Peccary. Ocelot has recently been recorded in that area. Cave Creek Canyon is the best place in North America to see White-nosed Coati; it also has Mexican and Allen's Big-eared Bats, American Black Bear (in summer), Striped and Hooded Skunks, Cougar, Bobcat, Ringtail, Collared Peccary, Mexican Fox Squirrel, Harris's Antelope Squirrel, Cliff Chipmunk, and Yellow-nosed Cotton Rat. Again, bird feeders are good places to look. Leslie Canyon National Wildlife Refuge has White-nosed Coati, White-tailed Deer of Coues's race, and rare Cockrum's Desert Shrew. Nearby **Chiricahua National Monument** has Western Bonneted Bat (look in vertical crevices in rock walls), Collared Peccary, Gray-collared Chipmunk, Cliff Chipmunk, Rock Squirrel of a distinctive black-faced race, Mexican Fox Squirrel, Yellow-nosed Cotton Rat, and Black-eared Mouse. Jaguar has been recorded in the Chiricahuas recently, but the chances of seeing one are practically zero.

The deserts around **Portal** (Arizona) are crossed by numerous roads good for night drives: almost all desert rodents of southern Arizona can be found here, including, for example, the Apache race of Plains Pocket Mouse and Tawny-bellied Cotton Rat. Look also for Bobcat and Collared Peccary.

The so-called "bootheel," the extreme southwestern corner of New Mexico, also has a few mostly Mexican species. Driving at night through the beautiful **Animas Valley** (along Highway 338), look for White-sided Jackrabbit as well as Crawford's Desert Shrew, Hooded Skunk, Kit Fox, and Merriam's and Banner-tailed Kangaroo Rats. Once you reach the mountain pass on the Arizona state line, look for Osgood's Mouse and (let's be optimistic!) Jaguar.

Although abundant in the East and the Midwest, White-Tailed Deer is rare and local in the West.

Apache-Sitgreaves National Forests on both sides of the Arizona/ New Mexico state line is the reintroduction site for Mexican Wolf. **Gila Cliff Dwellings National Monument**, located in the heart of this large plateau, is a good place to look for Southwestern Myotis, as well as for Osgood's Mouse, Northern Rock Mouse, and Mexican Vole.

The lowlands of central and southeastern New Mexico belong to another desert, the Chihuahuan, with its own distinctive fauna. The best mammal-viewing location in the southeastern part of the state is **Carlsbad Caverns National Park**, famous for a mass emergence of Mexican Free-tailed Bats every summer evening. If you can get up early enough, watch also the return of the bats at dawn; it's a very different experience. Fringed Myotis inhabits smaller caves in the area, while Ringtail, White-backed Hog-nosed Skunk, Striped Skunk, Banner-tailed and Ord's Kangaroo Rats, Silky and Hispid Pocket Mice, Northern Rock Mouse, and Mearns's Grasshopper Mouse are common around the caves and along the access road. Nearby **Sitting Bull Falls** are located at the end of a road that is very good for night drives; look for Cougar, Ringtail, and Nelson's Pocket Mouse. The city of **Carlsbad** is a reliable place to see Mexican Ground Squirrel, while the forests around the tiny settlement called **Sunspot** (15 miles from Cloudcroft) have lots of Gray-footed Chipmunks in summer.

Probably the most unusual desert location in New Mexico is **White Sands National Monument**. White gypsum sands in and around it are inhabited by pale-colored morphs of various desert mammals. Plains Pocket Mouse has evolved almost white coloration to match the environment. Look also for Pallid Bat (there used to be a colony in the visitor center), Desert Pocket Gopher, and introduced Gemsbok Oryx. Just a few hours' drive away, mammals living on black lava flows of central New Mexico have evolved very dark coloration. **El Malpais National Monument**, for example, has very dark-colored Stephens's Woodrat, Northern Rock Mouse, and Rock Squirrel. This park and the adjacent **El Malpais National Conservation Area** also have caves with Southwestern Myotis, Townsend's Big-eared Bat, and Mexican Free-tailed and Big Free-tailed Bats; you need to first obtain a permit to visit these caves. Abert's Squirrel and Cliff Chipmunk are common here.

The Rio Grande Valley, a deep geological rift, runs across two-thirds of New Mexico. Deserts of the valley floor have diverse mammal fauna that can be seen in **Sevilleta National Wildlife Refuge** and in Sandia Mountains foothills south of **Placitas**. Look for Texas Antelope Squirrel; Silky and Hispid Pocket Mice; Merriam's and Ord's Kangaroo Rats; Piñon, Deer, and White-footed Mice; and Southern Plains and Mexican Woodrats. **Bosque del Apache National Wildlife Refuge** in the middle of the valley has Black-tailed Prairie Dog, Banner-tailed Kangaroo Rat, Tawny-bellied Cotton Rat, and Meadow Jumping Mouse. The campus of New Mexico State University in the city of **Las Cruces** is a good place to look for Spotted and Mexican Ground Squirrels early in the morning (try desert scrub around the junction of Arrowhead Dr. and Wells St.).

Chains of isolated mountain ranges line the Rio Grande Valley on both sides. On the western side, the **Magdalena Mountains** are a good place to see Gray-collared Chipmunk, Rock Pocket Mouse, Cactus Mouse, and Bushy-tailed Woodrat. On the eastern side, the **Manzano** and possibly **Sandia Mountains** have the recently described Manzano Mountains Cottontail. Both ranges have also Rocky Mountain Elk (introduced), Snowshoe Hare, Rock Squirrel, Abert's Squirrel, the spruce race of Pine Squirrel, and Least and Colorado

Chipmunks. Conveniently, the summit of the Sandias (**Sandia Crest**) is accessible by road year-round (although the chipmunks are active only in May through October). The distinctive New Mexican race of Montane Shrew occurs in the **Capitan Mountains Wilderness** a bit farther east, and possibly on Sandia Crest as well.

Three particularly interesting sites are located in northwestern New Mexico. **Wheeler Peak**, the state's highest mountain, has animals typical to the Rocky Mountains of Colorado, such as American Pika, Golden-mantled Ground Squirrel, Yellow-bellied Marmot, and Western Heather Vole, all of which can be seen on a day hike to the summit. The remote **Chaco Culture National Historic Park**, famous for North America's most stunning Native American ruins, has White-tailed Antelope Squirrel, Plains (Apache race) and Silky Pocket Mice, Ord's and Banner-tailed Kangaroo Rats, Brush and Deer Mice, and Mexican and White-throated Woodrats. The most unique natural area is **Valles Caldera National Preserve**, a giant volcanic crater above Los Alamos. The grassy crater floor is only partially open to visitors; it is an excellent place to see Rocky Mountain Elk, American Badger, Coyote, Botta's Pocket Gopher, and Long-tailed Vole. Surrounding slopes are densely forested and inhabited by Montane and Dwarf Shrews, Northern Pocket Gopher, Deer Mouse, Southern Red-backed Vole, and Montane Vole.

Colorado Chipmunk is very common in the southern Rockies.

Northeastern New Mexico is part of the Great Plains. Mammals typical of shortgrass prairies, such as Pronghorn, Swift Fox, Least Chipmunk, Thirteen-lined Ground Squirrel, and Plains Pocket Gopher, can be seen around **Clayton Lake** and in **Kiowa National Grassland**.

The Great Plains part is a bit more extensive in Colorado. The best place to look for prairie mammals is **Comanche National Grassland**. In addition to the usual Pronghorn, American Badger, Swift Fox, Plains and Yellow-faced Pocket Gophers, Hispid Pocket Mouse, Northern Grasshopper Mouse, and Deer Mouse, it has interesting rodents living in deep, narrow canyons that cut across the prairie. This is probably the easiest place in North America to find Brush and Piñon Mice, and also tiny, very tame Silky Pocket Mouse. Abandoned houses on the prairie are often inhabited by Fringed Myotis, as well as Mexican and Southern Plains Woodrats.

The city of **Boulder** is one of the few places in the West where Black-tailed Prairie Dog can escape prosecution. Prairie dog towns are the best places to see American Badger, Red Fox, Coyote, Thirteen-lined Ground Squirrel, Northern Grasshopper Mouse, and Deer Mouse. Fields around Boulder have Hispid Pocket Mouse, Western Harvest Mouse, House Mouse, Prairie Vole, and Meadow Jumping Mouse, although the latter is very rare. Eastern Gray Squirrel has recently colonized city parks, while Abert's Squirrel and Uinta Chipmunk can be found on the slopes above the city. Rare Long-eared Myotis roosts in bird nest boxes in the area. In nearby **Denver**, wild Thirteen-lined Ground Squirrels live on the zoo grounds and are easier to see here than anywhere else.

The Colorado Rockies are formidable mountains where most roads to high elevations are closed in winter and spring. The road to **Guanella Pass** near Georgetown is usually the last to be closed. In years when the first snowfall is late or the snowmelt is early, you can drive up there and look for Long-tailed Weasel, Ermine, and Snowshoe Hare in white winter coats; they can be spotted from half a mile away against the dark background of snow-free tundra. Western Heather Vole also occurs there.

The highest road in North America climbs to the summit of **Mount Evans** (14,264 ft. [4,348 m]). It usually opens in early June and closes in September. Dwarf Shrew, Mountain Goat (introduced), Rocky Mountain Bighorn Sheep, American Pika, Colorado Chipmunk, Golden-mantled Ground Squirrel, and Yellow-bellied Marmot

Mountain Goats at Mount Evans.

are usually easy to see along the upper portions of the road. On summer nights, drivers are sometimes entertained by Coyotes that gorge themselves on thousands of moths attracted to the pavement by the car headlights (unfortunately, the road is no longer open at night, but you might be able to obtain a special permit). If you miss Bighorn Sheep on Mt. Evans, try **Bighorn Viewing Area** on I-70 near Vail.

Another road to alpine tundra is in **Rocky Mountain National Park**. Small mammals are similar there, but American Pika is more difficult to find. Rocky Mountain Elk and Wyoming Ground Squirrel are abundant in low-elevation parts of the park; Elk also graze above the timberline in summer. Rocky Mountain Bighorn Sheep, Cougar, and Plains Harvest Mouse are occasionally seen on the eastern side of the park, while Moose, Northern Pocket Gopher, and Beaver are common on the western side. American Marten can sometimes be spotted during hikes to high-elevation lakes.

The part of Colorado west from the Rockies is a bit similar to Utah. A large, flat plateau called **Grand Mesa** is a good place to see Montane Shrew and Northern Pocket Gopher in early spring, just as the meadows become snow-free. **Black Canyon of the Gunnison**

National Park to the south has Gunnison's Prairie Dog, as well as very dark-colored Canyon Mouse. Another good place for small mammals is **Great Sand Dunes National Park**, where Merriam's Shrew and Ord's Kangaroo Rat can be easily found by following their tracks in the sand at night or at dawn. Very scenic **Dinosaur National Monument** in the far northwestern corner of the state has Least Chipmunk, White-tailed Prairie Dog, and Olive-backed Pocket Mouse.

Colorado has a lot of caves, but they tend to be remote, and bat diversity is generally low. One easily accessible group of caves is in **Rifle Falls State Park**. The main bat colonies there are in caves with tiny entrances, but you can watch the emergence at dusk, and on weekdays sometimes see a few bats in larger caves. Little Brown Myotis is the most common species, but Long-eared Myotis, which is usually difficult to find in caves, is also present.

Texas and the Midwest

Texas is made up of two very different parts: the humid eastern half and the arid west. Both are intolerably hot in summer, but the heat is easier to tolerate in the west. Winters are mild, except during snowstorms, and the far south of the state doesn't have winters to speak of. Much of Texas has been gradually turning into desert in recent years thanks to the climate change (locally known as "the current drought") and the depletion of groundwater aquifers by nonsustainable agriculture.

Northern Texas and the Midwest are mostly a land of plains, gradually changing from shortgrass prairies in the west to tallgrass prairies (now almost completely plowed) to hardwood or coniferous forests in the east. Although the western part of the region is fairly high above sea level, the only mountains are the Chisos, Davis, and Guadalupe Mountains in far western Texas, the Black Hills in South Dakota, and the Ozark Plateau in the southeast. The climate of the Great Plains is not particularly pleasant, with sudden temperature changes, very hot summers, frequent snowstorms in winter and spring, and the world's most spectacular thunderstorms in spring and summer. For many people this violent weather is an attraction rather than a scourge, and you can expect to run into caravans of storm chasers' SUVs and minivans on local roads, particularly in April and May. Whether you like storm chasing or not, make sure your car insurance covers hail and tornado damage.

On the prairie, Northern Raccoon is mostly confined to riparian forests and towns. The ones in the far northern part of the range are the largest and most photogenic.

The entire region has more than 200 species of mammals, 20 of them marine. Of these, seven (Elliot's Short-tailed Shrew, three pocket gophers, two kangaroo rats, and Texas Mouse) are endemic, five are shared with only Mexico, and four with only New Mexico. Most animals of the Great Plains have extensive ranges, so recommending particular locations seldom makes sense. It is often better to look around your home than to waste time on long drives. There are, however, a few exceptions.

The most dramatic landscapes and the most diverse mammalian fauna are found in and around **Big Bend National Park**, where the forested Chisos Mountains rise from the desert just north of the Rio Grande. This cool "sky island" is one of the most likely places in North America to see Cougar. At night, American Black Bear, Collared Peccary, and Hooded Skunk wander around the Chisos Basin Campground, while Greater Long-nosed Bat visits flowering agave on the slopes (you can try to get a closer look at it by putting out a

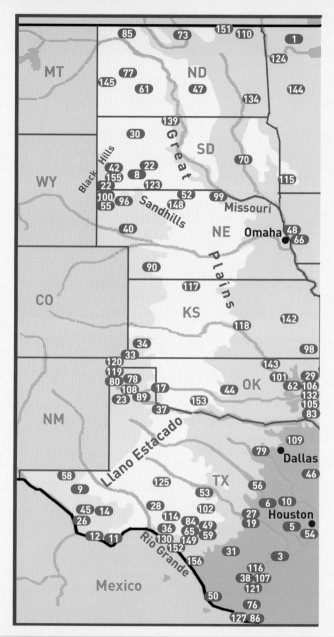

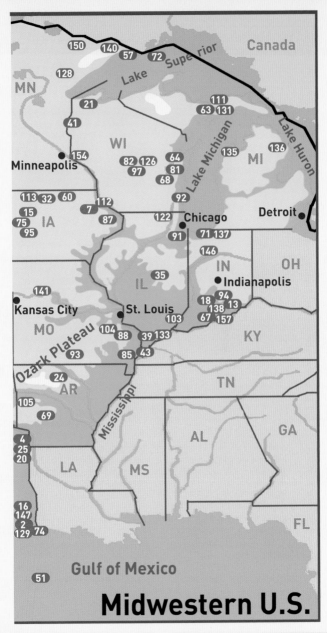

Midwestern U.S.

Texas and the Midwest

1. Agassiz NWR (MN)
2. Anahuac NWR (TX)
3. Aransas NWR (TX)
4. Atlanta SP (TX)
5. Attwater Prairie Chicken NWR (TX)
6. Austin (TX)
7. Backbone SP (IA)
8. Badlands NP (SD)
9. Balmorhea SP (TX)
10. Bastrop SP (TX)
11. Big Bend NP (TX)
12. Big Bend Ranch SP (TX)
13. Big Oaks NWR (IN)
14. Big Thicket NPR (TX)
15. Big Wall Lake (IA)
16. Black Gap WMA (TX)
17. Black Kettle NG (OK)
18. Bluespring Caverns (IN)
19. Bracken Bat Cave (TX)
20. Brandy Branch Reservoir (TX)
21. Brule River SF (WI)
22. Buffalo Gap NG (SD)
23. Buffalo Lake NWR (TX)
24. Buffalo National River (AR)
25. Caddo Lake SP (TX)
26. Capote Canyon (TX)
27. Cascade Caverns (TX)
28. Caverns of Sonora (TX)
29. Cherokee WMA (OK)
30. Cheyenne River Sioux Tribe (SD)
31. Choke Canyon SP (TX)
32. Clear Lake (IA)
33. Cimarron NG (KS)
34. Cimarron River (KS)
35. Clinton Lake SRA (IL)
36. Comstock (TX)
37. Copper Breaks SP (TX)
38. Corpus Christi (TX)
39. Crab Orchard NWR (IL)
40. Crescent Lake NWR (NE)
41. Crex Meadows WLA (WI)
42. Custer SP (SD)
43. Cypress Creek NWR (IL)
44. Darlington State Game Farm Reserve (OK)
45. Davis Mts. SP (TX)
46. Davy Crockett NF (TX)
47. Dawson WMA (ND)
48. DeSoto NWR (IA)
49. Eckert James River Bat Cave PR (TX)
50. Falcon SP (TX)
51. Flower Garden Banks NMS (TX)
52. Fort Niobrara NWR (NE)
53. Frio Cave (TX)
54. Galveston I. (TX)
55. Gilbert-Baker WMA (NE)
56. Gorman Cave (TX)
57. Grand Portage SF (MN)
58. Guadalupe Mts. NP (TX)
59. Haby Cave (TX)
60. Hayden Prairie SPR (IA)
61. Heart Butte Reservoir WMA (ND)
62. Heyburn WMA (OK)
63. Hiawatha NF (MI)
64. High Cliff SP (WI)
65. Hill Country SNA (TX)
66. Hitchcock NA (IA)
67. Hoosier NF (IN)
68. Horicon Marsh WLA (WI)
69. Hot Springs NP (AR)
70. Huron Wetland Management District (SD)
71. Indiana Dunes NLS (IN)
72. Isle Royale NP (MI)
73. J. Clark Salyer NWR (ND)
74. J. D. Murphree WMA (TX)
75. Kalsow Prairie SPR (IA)
76. Laguna Atascosa NWR (TX)
77. Lake Ilo NWR (ND)
78. Lake Meredith NRA (TX)
79. Lake Mineral Wells SP (TX)
80. Lake Rita Blanca (TX)
81. Ledge View Nature Center (WI)
82. Linnwood Springs Research Station (WI)
83. Little River NWR (OK)
84. Lost Maples SNA (TX)
85. Lostwood NWR (ND)
86. Lower Rio Grande Valley NWR (TX)
87. Maquoketa Caves SP (IA)

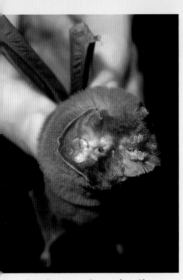

Frio Cave near Concan has Ghost-faced Bats.

88. Mark Twain NF (MO)
89. McClellan Creek NG (TX)
90. Medicine Creek SWA (NE)
91. Midewin National Tallgrass Prairie (IL)
92. Milwaukee (WI)
93. Mingo NWR (MO)
94. Miscatatuck NWR (IN)
95. Neal Smith NWR (IA)
96. Nebraska NF (NE)
97. Necedah NWR (WI)
98. Neosho WA (KS)
99. Niobrara SP (NE)
100. Oglala NG (NE)
101. Okmulgee WMA (OK)
102. Old Tunnel SP (TX)
103. Olney (IL)
104. Onondaga Cave (MO)
105. Ouachita Mts. (AR/OK)
106. Ozark Plateau NWR (OK)
107. Padre Island NSS (TX)
108. Palo Duro Canyon (TX)
109. Parkhill Prairie PR (TX)
110. Pembina Hills (ND)

111. Pictured Rocks NLS (MI)
112. Pikes Peak SP (IA)
113. Pilot Knob SP (IA)
114. Pinto (TX)
115. Pipestone NMT (MN)
116. Port Aransas (TX)
117. Prairie Dog SP (KS)
118. Quivira NWR (KS)
119. Rita Blanca NG (TX)
120. Rita Blanca WMA (OK)
121. Riviera Beach (TX)
122. Rock Cut SP (IL)
123. Rosebud Sioux Tribe (SD)
124. Rydell NWR (MN)
125. San Angelo (TX)
126. Sandhill WLA (WI)
127. Santa Ana NWR (TX)
128. Sax-Zim Bog (MN)
129. Sea Rim SP (TX)
130. Seminole Canyon SHP (TX)
131. Seney NWR (MI)
132. Sequoyah NWR (OK)
133. Shawnee NF (IL)
134. Sheyenne NG (ND)
135. Sleeping Bear Dunes NLS (MI)
136. South Fork Au Sable River (MI)
137. Spicer Lake Nature PR (IN)
138. Spring Mill SP (IN)
139. Standing Rock Sioux Tribe (SD)
140. Superior NF (MN)
141. Swan Lake NWR (MO)
142. Tallgrass Prairie NPR (KS)
143. Tallgrass Prairie PR (OK)
144. Tamarac NWR (MN)
145. Theodore Roosevelt NP (ND)
146. Tippecanoe River SP (IN)
147. Trinity River NWR (TX)
148. Valentine NWR (NE)
149. Valdina Farms Sinkhole (TX)
150. Voyageurs NP (MN)
151. Wakopa WMA (ND)
152. Webb Cave (TX)
153. Wichita Mts. NWR (OK)
154. Willow River SP (WI)
155. Wind Cave NP (SD)
156. Winter Haven (TX)
157. Wyandotte Caves (IN)

Eastern Pipistrelle is the smallest bat in the East, and the one most frequently encountered in eastern and midwestern caves.

hummingbird feeder). Other high-elevation species include Hoary Bat, Big Free-tailed Bat (there is a colony near the waterfall in Pine Canyon), Mule Deer, Carmen Mountain White-tailed Deer, the recently described Davis Mountains Cottontail, and Yellow-nosed Cotton Rat. At lower elevations, look for White-backed Hog-nosed Skunk, Pronghorn, Texas Antelope Squirrel, Rock Squirrel, Yellow-faced Pocket Gopher, Merriam's and Nelson's Pocket Mice, Merriam's Kangaroo Rat, Cactus and White-ankled Mice, Mearns's Grasshopper Mouse, and White-toothed Woodrat. Crevices in the rocky walls of Santa Elena Canyon shelter California Myotis, Pocketed Free-tailed Bat, and Western Bonneted Bat. The nearby **Big Bend Ranch State Park** has similar fauna, but Spotted Ground Squirrel and the Chihuahuan race of Desert Pocket Mouse are usually more common here.

Unlike the Chisos, the **Davis Mountains** to the north are mostly covered with grasslands. Local mammals include Collared Peccary, Davis Mountains Cottontail, Banner-tailed Kangaroo Rat, and Tawny-bellied and Yellow-nosed Cotton Rats. The highest mountains in Texas are the **Guadalupe Mountains**, shared with New Mexico. Gray-footed Chipmunk, the only chipmunk in Texas, occurs at the higher elevations there together with Rock Squirrel and Mexican Vole. Jones's Pocket Gopher is common in the foothills.

The northern Panhandle of Texas is a plain, slightly elevated and cut by deep canyons. The deepest of them, **Palo Duro Canyon**, is a good place to look for Ringtail, introduced Barbary Sheep, and the Comanche race of Piñon Mouse. **Rita Blanca National Grassland** and **Lake Rita Blanca State Park** in far northern Texas and adjacent **Rita Blanca Wildlife Management Area** in Oklahoma have many surviving Black-tailed Prairie Dog towns and a healthy population of Swift Fox. American Badger, Pronghorn, and Yellow-faced Pocket Gopher are also common here. Nearby **Lake Meredith National Recreation Area** is a good place to look for Striped Skunk and Northern Porcupine.

Although my favorite part of Texas is the far west, most people say that the most scenic area is the escarpment separating arid western Texas from the humid coastal plane. It is riddled with caves, some of which contain blind salamanders and fish, while others shelter humongous colonies of Mexican Free-tailed Bat. The world's largest bat colony, in **Bracken Bat Cave** near San Antonio, is 20 million strong;

White-tailed Deer populations in the Midwest were once actively managed by Native Americans.

the mass emergence of bats can be viewed a few times a week during warm months by taking a tour run by Bat Conservation International. Look for albinos among the emerging bats. In nearby **Old Tunnel State Park** and in **Eckert James River Bat Cave Preserve** near Mason, emerging Mexican Free-tails are joined by Cave Myotis bats. **Frio Cave** near Concan has Ghost-faced Bat and Cave Myotis in addition to lots of Mexican Free-tails. **Gorman Cave** near San Saba is open to visitors; it contains a colony of Cave Myotis, some Tricolored Bats, and occasionally an Eastern Red Bat. The prettiest cave in Texas, called **Caverns of Sonora**, has no bats, but Plains Pocket Gopher and Texas Mouse are common around the entrance. Other interesting places along the escarpment are **Bastrop State Park**, where Evening Bat is common in tree hollows; **Lost Maples State Natural Area**, where Llano Pocket Gopher occurs; and **Attwater Prairie Chicken National Wildlife Refuge**, a good place to see Plains Bison, Fulvous Harvest Mouse, and Thirteen-lined Ground Squirrel.

The far southern corner of Texas has a few remnant patches of dry tropical forest full of colorful birds. **Bentsen-Rio Grande Valley State Park** and nearby **Santa Ana** and **Lower Rio Grande Valley National Wildlife Refuges** have Crawford's Desert Shrew, Least Shrew, Bobcat, Ocelot, Collared Peccary, Mexican Ground Squirrel, Silky and Hispid Pocket Mice, Coues's Rice Rat, Ocelot, and possibly Jaguarundi. Look for Southern and Northern Yellow Bats roosting in dry palm fronds.

The coast of Texas is protected from frequent hurricanes by a chain of barrier islands. Numerous nature reserves line that coast. They can be a bit tough to explore in summer, but some are real wildlife heavens. The southernmost, **Laguna Atascosa National Wildlife Refuge**, is the only place in North America where seeing Ocelot is a real possibility. Jaguarundi is also said to be present, Bobcat is common, and Cougar has been recently recorded. Small mammals include Mexican Ground Squirrel, Texas Pocket Gopher, Hispid Cotton Rat, Mexican Spiny Pocket Mouse, Southern Plains Woodrat, Coues's Rice Rat, and the charming Northern Pygmy Mouse (look for the latter three around the observation blind and the bird feeders at the visitor center). The next one north, **Padre Island National Seashore**, has Gulf Coast Kangaroo Rat, and is the only place where Pantropical Spotted Dolphin is often seen from shore (elsewhere along the Gulf Coast you see only Common Bottlenose Dolphin). **Aransas National Wildlife Refuge** has Attwater's Pocket Gopher and Northern Pygmy

In Texas, Swamp Rabbit is most common in the humid, low-lying eastern parts of the state.

Mouse; Least Shrew is also common there. **Anahuac National Wildlife Refuge**, where nighttime access is allowed, has Baird's Pocket Gopher, Bobcat, American Mink, Marsh Rice Rat, and Swamp Rabbit. The nearby **Sea Rim State Park** is the best place in Texas to look for American Mink and introduced Coypu.

Flower Garden Banks National Marine Sanctuary, located far offshore southeast of Galveston, is visited by scuba diving trips a few times a year. These trips are expensive and often get cancelled because of bad weather, but they provide a rare opportunity to get far into the Gulf of Mexico, where Pantropical Spotted, Atlantic Spotted, Clymene, and Striped Dolphins occur, as well as Gervais's Beaked Whale and the offshore race of Common Bottlenose Dolphin.

The easternmost part of Texas is a land of dense hardwood forests and cypress swamps more typical to southeastern states. The most scenic swamps are in **Caddo Lake State Park**, which has Northern River Otter, Golden Mouse, American Beaver, Marsh Rice Rat, Hispid Cotton Rat, and Fulvous Harvest Mouse. The best hardwood forests are in **Big Thicket National Preserve**, where Southern Short-tailed Shrew, Fox Squirrel, and White-footed, Golden, and House Mice are common. Look for Seminole Bat feeding around streetlights at Kirby Trailhead and near the visitor center. On humid spring nights, when winged termites fly, Eastern Mole can sometimes be seen feeding on the ground around those lights. Plains Pocket Gopher occurs along Big Sandy Creek Trail.

Night driving on backcountry roads east of **Copper Breaks State Park** near Crowell in n.-cen. Texas is a good way to see the very rare Texas Kangaroo Rat. Texas and White-ankled Mice are also common there.

Oklahoma has a herd of reintroduced Plains Bison in **Wichita Mountains Wildlife Refuge**, which is also a good place to look for Elliot's Short-tailed Shrew, Least Shrew, Eastern Mole, Ringtail, Eastern Fox Squirrel, Hispid Pocket Mouse, Plains Harvest Mouse, Brush Mouse, and Eastern Woodrat. Eastern Small-footed Myotis, Texas Mouse, and many other small mammals are common in the **Ouachita Mountains** that stretch from Oklahoma into Arkansas.

Indiana, Missouri, and Arkansas have mammals typical for eastern woods, although there are a few prairie remnants here and there. All three states have plenty of caves; the most common bat species there are usually Tricolored and Big Brown Bats, sometimes also Little Brown Myotis. **Onondaga Cave** in Missouri has also Northern Myotis and sometimes Silver-haired Bat. In Indiana, caves of **Spring Mill State Park** have Eastern Small-footed, Indiana, and Southeastern Myotis, while Northern Short-tailed Shrew is common in the surrounding forest. **Wyandotte Caves** have Gray and Indiana Myotis; **Bluespring Caverns** have Northern Myotis. See species accounts in Part III for the best seasons to find them. **Hoosier National Forest**, also in Indiana, has lots of Hoary Bats, but they roost in trees rather than caves and are very difficult to find.

Cimarron National Grassland in far southwestern Kansas has many mammals of shortgrass prairies, such as Yellow-faced Pocket Gopher, Silky Pocket Mouse, Ord's Kangaroo Rat, and Northern Grasshopper Mouse. A very different environment is protected in **Tallgrass Prairie National Preserve** in eastern Kansas, home to Elliot's Short-tailed Shrew, Western Harvest Mouse, Hispid Cotton Rat, and very tame American Beaver.

Interesting places in Iowa include **Kalsow Prairie State Preserve** with Plains Pocket Gopher and Meadow Jumping Mouse; **Pilot Knob State Park**, where Southern Red-backed Vole occurs; **Clear Lake**, where Ermine is said to be particularly common; and two state parks

Black-tailed Prairie Dog is extirpated over large portions of the Midwest, but still common in the Texas Panhandle.

with a good variety of eastern forest mammals: **Pikes Peak** and **Backbone**.

The Nebraska Sandhills are an area of sandy soil where agriculture is difficult; that helped preserve large areas of grassland. For some reason, Eastern Spotted Skunk is more common and easier to see here than anywhere else. **Valentine National Wildlife Refuge** is a good place to look for it. Small mammals include Plains Pocket Mouse and Ord's Kangaroo Rat. Swift Fox still occurs in northwestern Nebraska, particularly in **Oglala National Grassland**.

South Dakota has two relatively small but beautiful national parks, **Wind Cave** and **Badlands**. Both have large colonies of Black-tailed Prairie Dog where Coyote and American Badger are often seen, as well as herds of Pronghorn, Mule Deer, and Plains Bison. The latter park has also Rocky Mountain Bighorn Sheep, Swift Fox, and reintroduced Black-footed Ferret (in Conata Basin on the southern border of the park), while the former is a good place to look for Prairie Shrew and Olive-backed Pocket Mouse. The nearby Black Hills have a good population of Cougar and a huge herd of Plains Bison in **Custer State Park**. **Theodore Roosevelt National Park** in North Dakota has similar fauna. The **Pembina Hills** in far northeastern North Dakota have Moose and Manitoba Elk. **Clark Salyer National Wildlife Refuge** is a good place for small grassland mammals, such as Western Jumping Mouse.

The northern parts of Minnesota, Wisconsin, and Michigan have boreal forests with northern mammals such as Arctic Shrew, Moose, Snowshoe Hare, and Northern Flying Squirrel. **Voyageurs National Park** in Minnesota and **Isle Royale National Park** in Michigan are particularly good for those, although some species are absent on Isle Royale. Voyageurs National Park is also a good place to look for Masked Shrew, Meadow and Rock Voles, Southern Red-backed Vole, and Least Chipmunk. The **Sax-Zim Bog**, also in Minnesota, has Gray Wolf, American Black Bear, American Marten, Moose, and American Beaver, although the carnivores are not easy to see there. **Grand Portage State Forest** in extreme northeastern Minnesota has Eastern Heather Vole, very rare in the U.S. **Willow River State Park** in Wisconsin has a mixture of forest and prairie species, including Masked Shrew and White-tailed Jackrabbit.

Cypress Creek National Wildlife Refuge in Illinois is an island of swamp habitat reminiscent of the Gulf Coast. Look for American Mink, American Beaver, Common Muskrat, and for numerous bats in hollow trees, particularly for rare Indiana Myotis.

Eastern U.S.

The East had probably been almost 100 percent forested prior to the arrival of the first humans. Forests are still extensive, but old-growth stands are now extremely rare, and forest composition has been altered almost everywhere. In the Southeast the forests are mostly hardwood, although there is a broad belt of upland pinewoods paral-

In summer, White-tailed Deer often graze in azalea groves on Appalachian summits.

lel to the Gulf Coast, extensive cypress swamps on floodplains, and other conifers in the mountains. In New England conifers gradually descend to lower elevations.

The Appalachian Mountains cut across the region, separating the coastal plain from the hills and plateaus of the interior. They are relatively low and don't create much of a climate difference. In the southern part of the region, winters are mild; springs and autumns are mostly pleasant and beautiful, especially in the Appalachians; but summers can be oppressively hot and humid—it takes most visitors a couple of weeks to get used to the heat. In the north the climate is much worse, with frequent storms and extended periods of bad weather. Southern Florida has a tropical climate, although winter temperatures do fall below freezing once every few years. June to December is hurricane season; hurricane-caused floods might provide a unique opportunity to see some cryptic mammals such as Eastern Mole and various shrews.

Despite its lush forests and high overall biodiversity, the East has

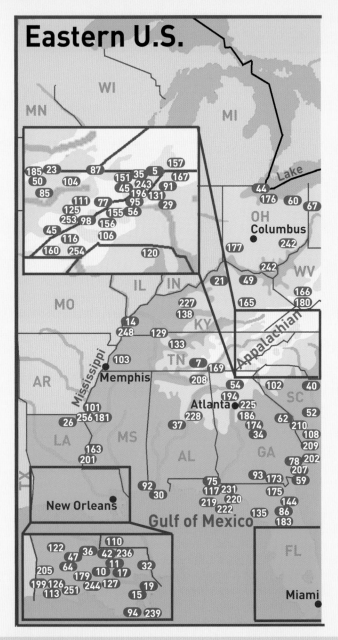

Eastern U.S.

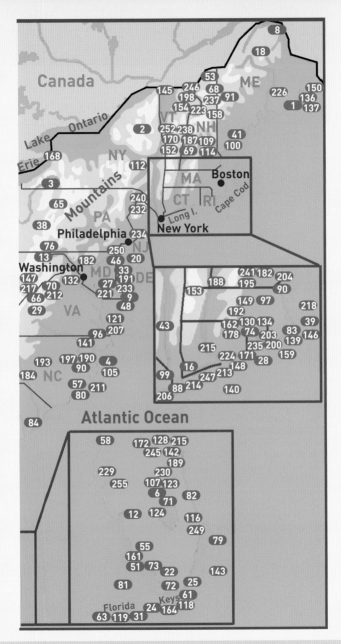

Eastern U.S.

1. Acadia NP (ME)
2. Adirondack Mts. (NY)
3. Allegany SP (NY)
4. Alligator River NWR (NC)
5. Appalachian Caverns (TN)
6. Archbold Biological Station (FL)
7. Arnold Air Force Base (TN)
8. Aroostook NWR (ME)
9. Assateague I. (VA)
10. Atchafalaya NWR (LA)
11. Audubon SHS (LA)
12. Babcock Webb WMA (FL)
13. Backbone Mt. (MD/WV)
14. Ballard WMA (KY)
15. Barataria PR (LA)
16. Bartlett Arboretum (CT)
17. Baton Rouge (LA)
18. Baxter SP (ME)
19. Bayou Savage NWR (LA)
20. Belleplain SF (NJ)
21. Bernheim Arboretum and Research Forest (KY)
22. Big Cypress NPR (FL)
23. Big South Fork NRA (KY/TN)
24. Big Torch Key (FL)
25. Bill Baggs Cape Florida SP (FL)
26. Black Bayou Lake NWR (LA)
27. Blackwater NWR (MD)
28. Block I. (RI)
29. Blue Ridge Pkwy. (NC/VA)
30. Bon Secour NWR (AL)
31. Boca Chica Key (FL)
32. Bogue Chitto NWR (LA)
33. Bombay Hook NWR (DE)
34. Bond Swamp NWR (GA)
35. Bristol Caverns (TN)
36. Bunkie (LA)
37. Cahaba River NWR (AL)
38. Canoe Creek SP (PA)
39. Cape Cod NSS (MA)
40. Carolina Sandhills NWR (SC)
41. Casco Bay (ME)
42. Cat Island NWR (LA)
43. Catskill Mts. (NY)
44. Cedar Point NWR (OH)
45. Cherokee NF (TN)
46. Chesapeake and Delaware Canal WA (DE)
47. Chicot SP (LA)
48. Chincoteague NWR (VA)
49. Clay WMA (KY)
50. Colditz Cove SNA (TN)
51. Collier Seminole SP (FL)
52. Congaree NP (SC)
53. Connecticut Lakes SF (NH)
54. Coosawattee WMA (GA)
55. Corkscrew Swamp Sanctuary (FL)
56. Craggy Dome (NC)
57. Croatan NF (NC)
58. Crystal River (FL)
59. Cumberland Island NSS (GA)
60. Cuyahoga Valley NP (OH)
61. Dagny Johnson Key Largo Hammock Botanical SP (FL)
62. Di-Lane Plantation WMA (GA)
63. Dry Tortugas NP (FL)
64. Duralde Cajun Prairie (LA)
65. Elk Country Visitor Center (PA)
66. Endless Caverns (VA)
67. Erie NWR (PA)
68. Errol (NH)
69. Esther Currier WMA (NH)
70. Evans Tract (VA)
71. Everglades Headwaters NWR (FL)
72. Everglades NP (FL)
73. Fakahatchee Strand PR SP (FL)
74. Fisherville Brook WR (RI)
75. Florida Caverns SP (FL)
76. Forbes SF (PA)
77. Forbidden Caverns (TN)
78. Fort Frederica NMT (GA)
79. Fort Lauderdale–Bahamas ferries (FL)
80. Fort Mason NP (NC)
81. Fort Myers–Key West ferry (FL)
82. Fort Pierce (FL)
83. Francis A. Crane WMA (MA)
84. Francis Marion NF (SC)
85. Frozen Head SP (TN)
86. Gainesville (FL)
87. Gap Cave; Cumberland Gap NHP (KY/TN/VA)
88. Gateway NRA (NJ/NY)
89. Gloucester (MA)

90. Goose Creek SP (NC)
91. Grafton Notch SP (ME)
92. Grand Bay NWR (MS)
93. Grand Bay WMA (GA)
94. Grand Isle (LA)
95. Grandfather Mt. (NC)
96. Great Dismal Swamp NWR (NC/VA)
97. Great Meadows NWR (MA)
98. Great Smoky Mts. National Park (NC/TN)
99. Great Swamp NWR (NJ)
100. Green I. (ME)
101. Greenville (MS)
102. Greenwood (SC)
103. Hatchie NWR (TN)
104. Hatfield Knob (TN)
105. Hatteras (NC)
106. Hickory Nut Gorge (NC)
107. Highland Hammock SP (FL)
108. Hilton Head I. (SC)
109. Hoit Rd. Marsh WMA (NH)
110. Homochitto NF (MS)
111. House Mt. SNA (TN)
112. Howe Caverns (NY)
113. Hwy. 82 (LA)
114. John Hay NWR (NH)
115. Jonathan Dickinson SP (FL)
116. Joyce Kilmer Memorial Forest (NC)
117. Judges Cave (FL)
118. Key Largo (FL)
119. Key West (FL)
120. Kings Mt. SP (SC)
121. Kiptopeke SP (VA)
122. Kisatchie NF (LA)
123. Kissimmee Prairie Preserve SP (FL)
124. Kissimmee River Public Use Area (FL)
125. Knoxville (TN)
126. Lacassine NWR (LA)
127. Lake Fausse Pointe SP (LA)
128. Lake Woodruff NWR (FL)
129. Land Between The Lakes NRA (KY/TN)
130. Lime Rock PR (RI)
131. Linville Gorge (NC)
132. Long Branch Nature Center (VA)
133. Long Hunter SP (TN)
134. Lonsdale Marsh (RI)
135. Lower Suwannee NWR (FL)
136. Machias Bay (ME)
137. Machias Seal I. (ME)
138. Mammoth Cave NP (KY)
139. Mashpee NWR (MA)
140. McMaster Submarine Canyon
141. Merchants Mill Pond SP (NC)
142. Merritt Island NWR (FL)
143. Miami–Nassau ferries (FL)
144. Mike Roess Gold Head Branch SP (FL)
145. Missisquoi NWR (VT)
146. Monomoy NWR (MA)
147. Monongahela NF; Dolly Sods (WV)
148. Montauk Pt. SP (NY)
149. Moore SP (MA)
150. Moosehorn NWR (ME)
151. Morrill Cave (TN)
152. Morris Cave (VT)
153. Mt. Greylock (MA)
154. Mt. Mansfield (VT)
155. Mt. Mitchell SP (NC)
156. Mt. Pisgah (NC)
157. Mt. Rogers (VA)
158. Mt. Washington (NH)
159. Muskeget I. (MA)
160. Nantahala NF (NC)
161. Naples (FL)
162. Natchaug SF (CT)
163. Natchez (MS)
164. National Key Deer Refuge (FL)
165. Natural Bridge SP (KY)
166. New River Gorge (WV)
167. New River SP (NC)
168. Niagara Falls SP (NY)
169. Nickajack Cave (TN)
170. Nickwackett Cave (VT)
171. Ninigret NWR (RI)
172. Ocala NF (FL)
173. Okefenokee NWR (GA)
174. Okmulgee NMT (GA)
175. Osceola NF (FL)
176. Ottawa NWR (OH)
177. Oxford (OH)
178. Pachaug SF (CT)
179. Palmetto Island SP (LA)
180. Panther WMA (WV)
181. Panther Swamp NWR (MS)

182. Patuxent Research Refuge (MD)
183. Paynes Prairie PR SP (FL)
184. Pee Dee NWR (NC)
185. Pickett SP (TN)
186. Piedmont NWR (GA)
187. Pillsbury SP (NH)
188. Pisgah SP (NH)
189. Platt Branch WMA (FL)
190. Pocosin Lakes NWR (NC)
191. Prime Hook NWR (DE)
192. Purgatory Chasm State Reservation (MA)
193. Raven Rock SP (NC)
194. Red Top Mt. SP (GA)
195. Rhododendron SP (NH)
196. Roan Mt. SP (TN)
197. Roanoke River NWR (NC)
198. Rte. 114 (VT)
199. Sabine NWR (LA)
200. Sachuest Point NWR (RI)
201. St. Catherine Creek NWR (MS)
202. St. Simons I. (GA)
203. Sakonnet Pt. (RI)
204. Salisbury Beach (MA)
205. Samuel Houston Jones SP (LA)
206. Sandy Hook (NJ)
207. Sapelo I. (GA)
208. Sauta Cave NWR (AL)
209. Savannah NWR (SC)
210. Savannah River Site (SC)
211. Shackleford Banks (NC)
212. Shenandoah NP (VA)
213. Shinnecock Bay (NY)
214. Silver Point CP (NY)
215. Sleeping Giant SP (CT)
216. Smyrna Dunes Park (FL)
217. Spruce Knob (WV)
218. Stellwagen Bank NMS (MA)
219. St. Joseph Peninsula SP (FL)
220. St. Marks NWR (FL)
221. St. Mary's River SP (MD)
222. St. Vincent NWR (FL)
223. Steam Mill Brook WMA (VT)
224. Stewart B. McKinney NWR (CT)
225. Stone Mt. (GA)
226. Sunkhaze Meadows NWR (ME)
227. Swan Lake WMA (KY)
228. Talladega NF (AL)
229. Tampa (FL)
230. Three Lakes WMA (FL)
231. Torreya SP (FL)
232. Tranquility (NJ)
233. Trap Pond SP (DE)
234. Trenton (NJ)
235. Trustom Pond NWR (RI)
236. Tunica Hills (LA)
237. Umbagog NWR (ME/NH)
238. Union Village (VT)
239. Venice (LA)
240. Wallkill River NWR (NJ)
241. Wapack NWR (NH)
242. Wayne NF (OH)
243. Weaver's Bend (TN)
244. Weeks I. (LA)
245. Wekiwa Springs SP (FL)
246. Wenlock WMA (VT)
247. Wertheim NWR (NY)
248. West Kentucky WMA (KY)
249. West Palm Beach (FL)
250. White Clay Creek SP (DE/PA)
251. White Lake WCA (LA)
252. White Rocks NRA (VT)
253. Whites Mill Refuge (TN)
254. Whiteside Mt. (NC)
255. Withlacoochee SF (FL)
256. Yazoo NWR (MS)

Northern Raccoons can turn any trip to a suburban park into a jungle experience.

only about 160 species of mammals, 35 of them marine. At least eight species (Florida Bonneted Bat, Marsh Rabbit, New England Cottontail, Southeastern Pocket Gopher, two mice, Appalachian Woodrat, and Round-tailed Muskrat) are endemic to the region, and a few bat species are shared with only the West Indies. Except for a few abundant species, mammals are much more difficult to find than in the West. It's not uncommon to spend a whole night walking or driving on forest roads and see nothing but White-tailed Deer, Northern Raccoon, and Virginia Opossum. But on a good night, up to seven species can sometimes be found.

The **Florida Keys** are the most "tropical" place in North America. These small islands have distinctive races of many mammal species. The tiny Key Deer, which is easy to see on Big Pine Key, is the best-known one. There are also island races of Marsh Rice Rat on Big Torch Key, of Marsh Rabbit and Northern Raccoon on both Key Largo and Big Pine Key, and of Cotton Mouse and Eastern Woodrat on Key Largo. The best place to look for mammals on Key Largo is the terribly named **Dagny Johnson Key Largo Hammock Botanical State Park**. Big Pine Key and adjacent islands are protected as **Key Deer National Wildlife Refuge**. Two West Indian bat species occur in North America only on the Keys: Velvety Free-tailed Bat is common in **Marathon**, while Jamaican Fruit-eating Bat is rare in **Key West**. Three more West Indian bat species have been recorded on the Keys only once or twice.

Ferries go from Key West to Fort Myers and the Dry Tortugas. Most of the time only Common Bottlenose Dolphin is seen from those ferries, but Long-beaked Common, Spinner, and Atlantic Spotted Dolphins have also been observed. Other ferries go from **Fort Lauderdale** to Freeport in the Bahamas, and from Miami to Nassau. Although marine mammals are rarely seen during these crossings, Bryde's and Humpback Whales, and Pantropical Spotted, Striped, and Short-beaked Common Dolphins can occasionally be spotted, and many other species such as Dwarf Sperm Whale are theoretically possible.

Everglades National Park used to be one of the best mammal-watching sites in eastern North America, but recently many small mammals have virtually disappeared, probably as a result of Burmese python and Argentine fire ant introductions. Cotton Mouse and the distinctive Everglades race of Southern Short-tailed Shrew can still be seen in Mahogany Hammock. Others include the endemic Florida

Bonneted Bat (along Pineland Trail), Seminole Bat (in trees growing in the Anhinga Trail parking lot), Florida Panther (near the main park entrance and in the Hole-in-the-Donut area), Round-tailed Muskrat and the rare Everglades race of American Mink (along the eastern side of Shark Valley Loop), and Common Bottlenose Dolphin and Caribbean Manatee (at the entrance to the Flamingo harbor). **Big Cypress National Preserve**, **Corkscrew Swamp Sanctuary**, and **Fakahatchee Strand Preserve State Park** are much better for mammal watching nowadays. There you can see lots of common animals such as Northern River Otter, Northern Raccoon, Virginia Opossum, and Marsh Rabbit, as well as less common ones such as Least Shrew, the Florida race of American Black Bear (try Big Cypress Bend Trail or Bear Island), the Mangrove race of Fox Squirrel (try Loop Road), and sometimes Florida Panther (look along Turner River Road). **Collier-Seminole State Park** has the distinctive Sherman's race of Southern Short-tailed Shrew. If you are interested in just mammals and not in reptiles, it is better to visit southern Florida between October and early May, because in summer there is extensive flooding, and biting insects can be so plentiful that they pose a serious survival risk in deep woods.

In central Florida the most interesting mammals, some of them unique to the peninsula, live in the so-called sand scrub habitat, which exists in small patches at **Archbold Biological Station** near Lake Placid, in **Wekiwa Springs State Park**, and in the Big Scrub area of **Ocala National Forest**. Look for Northern Yellow Bat, Sherman's race of Fox Squirrel, Southeastern Pocket Gopher, and Oldfield and Florida Mice in these places. Archbold Biological Station also has Mexican Free-tailed Bat, Southern Flying Squirrel, and Jaguarundi (possibly introduced). Good knowledge of animal tracks will help you greatly in this sandy habitat.

Highlands Hammock State Park has an outstanding collection of various forest types found in central Florida, from hardwood forests to pinelands to cypress swamps. In addition to ubiquitous Virginia Opossum, Northern Raccoon, and Nine-banded Armadillo, it has the Everglades race of Southern Short-tailed Shrew, Seminole Bat (look along the cypress swamp boardwalk), Bobcat, Cotton Mouse, Hispid Cotton Rat, and Eastern Woodrat. **Three Lakes Wildlife Management Area** and **Kissimmee Prairie Preserve State Park** are good places to look for Eastern Spotted Skunk, Coyote, Marsh Rabbit, Fox Squirrel of Sherman's race, and Feral Pig. **Canaveral National Sea-**

Southern Flying Squirrel can be found by spotlighting in tall forests of the East.

shore and **Smyrna Dunes Park** on the east coast have beautiful white Beach Mouse and Eastern Spotted Skunk (but both places are technically closed at night). In summer you can sometimes swim with Caribbean Manatees at Canaveral National Seashore, but in winter it is better to cross the peninsula to the west coast and swim with them in **Crystal River**, where manatee-viewing tours are offered. **Lake Woodruff National Wildlife Refuge** is the most reliable place to see Round-tailed Muskrat; it also has lots of Marsh Rabbits.

In northern Florida, **Lower Suwannee National Wildlife Refuge** has a distinctive race of Meadow Vole. **Judges Cave** near Marianna has Gray and Southeastern Myotis (the cave is gated, but you can see the bats emerge at dusk), while **Torreya State Park** is a good place to look for Golden Mouse. **St. Vincent National Wildlife Refuge**, located on the island of the same name, has a few reintroduced Red Wolves, but the chances of seeing one are slim unless you manage to obtain an overnight stay permit. There are also introduced Sambar Deer on the island.

Just across the Georgia state line, **Okefenokee National Wildlife Refuge** is another good area to look for Southern Short-tailed Shrew, Northern River Otter, Striped Skunk, and Marsh Rice Rat. **Bond Swamp National Wildlife Refuge**, also in Georgia, has remarkably diverse habitats and lots of interesting mammals, including Southeastern and Least Shrews, Seminole and Silver-haired Bats, Eastern Spotted Skunk, American Black Bear, Swamp and Marsh Rabbits, Southeastern Pocket Gopher, Eastern Harvest Mouse, Eastern Woodrat, Woodland Vole, and Cotton, Oldfield, and Golden Mice. An interesting place on the coastal plain is the **Savannah River Site** in South Carolina, near Augusta. It is closed to the public, but if you drive through along Highway 125, you have a very good chance of

A swamp-living American Black Bear, Florida.

seeing Bobcat, particularly at dusk. Carolina Dog still occurred there a few years ago. Farther north, **Alligator River National Wildlife Refuge** in North Carolina is a huge area of swamps and impenetrable forests, where Eastern Red Bat and American Black Bear are very common. Red Wolf has been reintroduced here but is difficult to see (it took me more than a week). Even farther north, **Great Dismal Swamp National Wildlife Refuge** on the North Carolina/Virginia state line has Northern Short-tailed and Southeastern Shrews, Marsh Rabbit, and very tame American Beaver.

Hatteras in North Carolina is the only place in the Southeast where bird-watching tours into the open ocean are regularly available (run by **Seabirding Pelagic Trips**). Marine mammals are a bit sparse, but Common Bottlenose (inshore and offshore forms), Atlantic Spotted, Spinner, Risso's, and Long-beaked Common Dolphins, as well as Humpback and Sperm Whales, are often seen. False Killer Whale, Short-finned and Long-finned Pilot Whales, and Cuvier's and Gervais's Beaked Whales have also been recorded, and True's Beaked Whale is possible.

Long Branch Nature Center in Arlington (Virginia) sets up a feeding program for Southern Flying Squirrels every winter, providing a rare chance to see these nocturnal animals up close. They usually show up about an hour after sunset.

The Gulf Coast has some beautiful wilderness, but you have to be a really dedicated naturalist to enjoy it in summer, thanks to frequent floods, clouds of biting insects, and the catastrophic infestation by introduced Argentine fire ant ("the South's most hated invader since general Sherman," as one local newspaper put it). It's a tough place to live for a mammal, and mammal watching here is seldom very productive, in sharp contrast with the fantastic bonanza bestowed upon local herpetologists and ichthyologists. Good places to see forest fauna include **Cat Island**, **Sabine** and **Atchafalaya National Wildlife Refuges** in Louisiana, and **St. Catherine Creek National Wildlife Refuge** in Mississippi. Look for Seminole and Evening Bats, Swamp Rabbit, Fulvous and Eastern Harvest Mice, and Cotton Mouse. Northern Raccoon, Virginia Opossum, Nine-banded Armadillo, and introduced Coypu are abundant. In rice- and crawfish-growing areas of Louisiana, Southern Short-tailed and Least Shrews, American Mink, Marsh Rice Rat, and Fulvous Harvest Mouse are very common. One of the best places to look for them is **Duralde Cajun Prairie** near Eunice (Louisiana), located at 30°33'31" N, 92°29'10" W. In forested areas, Rafinesque's Big-eared Bat can often be found in hollow tupelo trees and under road bridges. In summer, mosquito densities in floodplain forests of the Gulf Coast are so high that some nocturnal mammals become active in early mornings. **Lake Fausse Pointe State Park** in Louisiana has a good network of muddy trails where Nine-banded Armadillo, Louisiana Black Bear, and McIlhenny's race of White-tailed Deer can often be found. Eastern Cougar has been seen in that park recently.

There is no easy way to get far out into the Gulf of Mexico from Louisiana, Mississippi, or Alabama. The cheapest option for getting to deep waters is to hitch a ride on a fishing boat heading south from **Grand Isle** or **Venice**, Louisiana. Many species of whales and dolphins, including some very rare ones, have been recorded in this area, but population densities are generally low and nothing is guaranteed; the inshore form of Common Bottlenose Dolphin is the only species seen regularly.

The Appalachian Mountains are much more welcoming than the coastal plains. Even a hopeless city dweller would enjoy a hike here, particularly during the spring bloom or the brilliant fall colors season. There are two beautiful national parks in the southern Appalachians:

Great Smoky Mountains and **Shenandoah**. They are connected by the **Blue Ridge Parkway**, which is really good for night drives. The better of the two is Great Smoky Mountains National Park. Its most popular part, Cades Cove, can be comfortably visited only at dawn; later it turns into one huge traffic jam. Cades Cove meadows are frequented by Pygmy Shrew (along streams), Coyote, American Black Bear (mostly on the southern side), White-tailed Deer, Eastern Harvest Mouse, and Meadow Vole. In hardwood forests, look for Smoky Shrew, Northern Short-tailed Shrew, Eastern Red and Hoary Bats, the Appalachian race of New England Cottontail, White-footed and Deer Mice, and Appalachian Woodrat. Little River, Laurel Creek, and Big Cove Roads are good for spotlighting after midnight. Elk has been reintroduced to the North Carolina part of the park, and is often seen near the visitor center on the southern side (also a good place for Woodchuck). Rafinesque's Big-eared Bat occurs in log cabins in summer. The highest peak in the park, Clingmans Dome, is accessed by a paved road that is closed in winter. It usually becomes snow-free two to three weeks before opening for traffic, so you can walk to the summit in perfect solitude and look for Rock Shrew, Northern Flying Squirrel, Southern Red-backed Vole, and Rock Vole. **Roan Mountain State Park** in Tennessee, famous for its rhododendron blooms, and Gregory Bald in Great Smoky Mountains National Park, famous for its azalea blooms, are particularly good places to look for Hairy-tailed Mole, because the animals often cannot dig through the tangled azalea or rhododendron roots and have to crawl on the surface. At low-elevation parts of the Blue Ridge Parkway, look for Gray Fox, Striped Skunk, Eastern Spotted Skunk, Fox Squirrel, Southern Flying Squirrel, and Woodchuck. **Mount Mitchell**, **Craggy Dome**, and **Mount Pisgah** in North Carolina, **Mount Rogers** in Virginia, and **Spruce Knob** in West Virginia have fauna similar to that of Clingmans Dome.

West of the Appalachians lies the Cumberland Plateau. Scattered among its beautiful valleys are places where rocks form huge overhangs, walls, and niches. **Colditz Cove State Natural Area** in Tennessee is one of the best such places, inhabited by Smoky and Southeastern Shrews, Eastern Spotted Skunk, Cotton Mouse, Eastern Woodrat, and Woodland Jumping Mouse. Another such place is **Natural Bridge State Resort Park** in Kentucky; it has Rafinesque's and Townsend's Big-eared Bats, but seeing them is difficult. **Frozen Head State Park** in Tennessee is a good place to look for Eastern Mole,

Introduced Manitoban Elk at Hatfield Knob, Tennessee.

Eastern Chipmunk, and Woodland Vole. **Hatfield Knob**, a mountain near La Follett in Tennessee, has a reintroduced Elk herd and a watchtower. You can watch elk rut in the fall, or simply observe the clearing. White-tailed Deer, Striped Skunk, Coyote, and Bobcat often show up. Cottontail Rabbits seen along the access road might belong to an undescribed species. A bit to the north, in the area where Tennessee, Kentucky, and Virginia meet, is another interesting place, **Cumberland Gap National Historic Park**, where Smoky Shrew and Golden Mouse occur. Even farther north, **Bernheim Arboretum and Research Forest** in Kentucky has Southern Bog Lemming.

Eastern Chipmunk is abundant in many forests of the East, but for some reason it is relatively rare in Great Smoky Mountains National Park.

The western side of the Appalachians and the Cumberland Plateau have a lot of caves. Almost every cave in the region has Tricolored and Big Brown Bats in winter. **Sauta Cave National Wildlife Refuge** in Alabama has the largest colony of Gray Myotis; although the cave is gated, in summer and early fall you can watch the bats fly out at dusk. **Gap Cave** in Cumberland Gap National Historic Park often has Indiana and Little Brown Myotis. One of the few caves you can explore by yourself is **Morrill Cave** (36.46°N, 82.23°W) near Bluff City in Tennessee. If you fill out the self-registration form and pay the parking fee at the owner's house, you can walk straight into the cave, which has Northern and Little Brown Myotis, Tricolored Bat, and Big Brown Bat in winter. The largest cave (by far the world's longest) is **Mammoth Cave** in Kentucky, where Gray and Indiana Myotis are often visible during tours. Eastern Woodrat and Woodland Vole occur in the woods around the cave, while Eastern Harvest Mouse lives in forest clearings there.

In Pennsylvania, **Elk Country Visitor Center** near Benezette is a great place to see Elk rut. American Black Bear, Bobcat, and Fisher also occur in that area and can potentially be seen from the observation tower. **Wayne National Forest** in Ohio has good habitat diversity and most mammal species of the eastern woods.

Mammal watching in the northeastern U.S. is a bit different from the rest of the continent. This area is highly urbanized, and its fauna is impoverished by recent glaciation and the all-out extermination of wildlife following the European colonization, although some species such as Fisher and Northern River Otter are slowly coming back now. Many protected natural areas are very small, but that doesn't mean they are not interesting; they are just difficult to learn about if you don't know at least their names in advance.

One such little gem is the **Esther Currier Wildlife Management Area** in New Hampshire, an excellent place to see Ermine, Southern Red-backed Vole, Woodland Vole, and American Beaver. With some luck, you can also find Star-nosed Mole, American Marten, Fisher, Mink, and Meadow Vole. Hidden in plain sight in the city of Pawtucket (Rhode Island) is **Lonsdale Marsh**, a good place to find Northern Short-tailed Shrew, Hairy-tailed Mole, Southern Red-backed Vole, and Meadow Jumping Mouse along the numerous paths leading to the river. **Lime Rock Preserve**, also in Rhode Island, has Woodland Vole, nonferal Woodchuck, Deer Mouse, and Least Weasel. In New Jersey, **Belleplain State Forest** has the distinctive Maryland

Northern Myotis used to be common in northeastern caves, but is now threatened with extinction from white-nose syndrome.

race of Masked Shrew, while **Wallkill River National Wildlife Refuge** has large predators such as American Black Bear and Bobcat. **Great Swamp National Wildlife Refuge**, located in New Jersey just a few miles from downtown Manhattan, is reportedly one of the best places to see Star-nosed Mole; it is also good for Virginia Opossum, Northern Short-tailed Shrew, Northern Raccoon, Northern River Otter, Woodland Vole, and Meadow Jumping Mouse. The rarest small mammal in the Northeast is New England Cottontail. Look for it in the Salt Meadow Unit of **Stewart B. McKinney National Wildlife Refuge** (Connecticut), in and around **Mashpee National Wildlife Refuge** in the southern part of **Cape Cod**, along any trail through dense brush in **Cape Cod National Seashore** (both in Massachusetts), and in **Ninigret National Wildlife Refuge** in Rhode Island. The latter place also has Southern Bog Lemming. Yet another good place in Rhode Island is **Trustom Pond National Wildlife Refuge**, one of the best places to see Southern Red-backed Vole (along Red Maple Swamp Trail), tame Eastern Chipmunk (around the bird feeders near the visitor center), and American Mink (along rocky shores at low tide).

As you move north, the fauna begins to change. By the time you reach northern New England, Moose becomes common (see Part III for a list of viewing locations), and Snowshoe Hare replaces Eastern Cottontail. The extensive forests of the **Adirondack Mountains** in New York State have a blackish race of Snowshoe Hare, as well as numerous small mammals, including Smoky Shrew and Woodland Jumping Mouse. High mountains such as **Mount Mansfield** in Vermont, **Mount Washington** in New Hampshire, and the highlands of **Baxter State Park** in Maine are usually the best places to see northern species. Other interesting sites in New Hampshire are **Pisgah State Park**, where Fisher occurs, and **Pillsbury State Park**, where Northern River Otter and American Mink are said to be relatively easy to see. Another good place for aquatic mammals such as Star-nosed Mole, American Mink, American Beaver, and Common Muskrat is **Missisquoi National Wildlife Refuge** in Vermont. In Maine, **Acadia National Park** has almost all mammalian species of the area and an excellent network of trails to look for them. Unlike most other protected natural areas in New England, it is open at night. Star-nosed Mole, the weirdest-looking mammal on the continent, is said to be particularly common in **Sunkhaze Meadows National Wildlife Refuge**, also in Maine.

The large **Muskeget Island** race of Meadow Vole lives on its namesake island, accessible by a somewhat risky sea kayaking trip from Nantucket. Muskeget and **Monomoy Islands** have the largest breeding colonies of Gray Seal in the U.S. Gray Seal can also be seen in Acadia National Park year-round. In winter it shows up farther south, all the way to Rhode Island, and is joined by Harbor Seal, which is usually more numerous in winter and occurs south to New Jersey. **Silver Point County Park** and **Montauk Point State Park** in New York and **Sandy Hook** in New Jersey are good places to see Harbor Seal in winter.

Whale-watching tours are run in summer from Montauk Point; from **Gloucester** (Massachusetts), where **Cape Ann Whale Watch** gets particularly good reviews; and from **Provincetown** at the tip of Cape Cod. Many of these tours go to **Stellwagen Bank National Marine Sanctuary**. Regularly seen species here include Northern Minke, Fin, Humpback, and (in early May) Northern Right Whales, as well as Short-beaked Common Dolphin and (in spring and fall) Atlantic White-sided Dolphin. Occasionally, Long-finned Pilot Whale, Sei Whale, and White-beaked Dolphin are encountered.

Eastern Canada, Nunavut, and Greenland

Compared to the northeastern U.S., eastern Canada is surprisingly pristine, and seems almost unpopulated. As you travel north, roads rapidly become sparse and then disappear. The Arctic wilderness of northern Labrador, Nunavut, and Greenland is teeming with fascinating wildlife, but travel there is so expensive that few people can afford it.

Summers are cool and wet in mainland Canada, but dry in Nunavut and Greenland; Ellesmere Island is drier than much of the Sahara. Winters are long and harsh. Fall colors in the southern part of the region are the world's best, especially in southwestern Québec. In the north, global warming is beginning to destroy natural habitats, but thanks to extensive ice plateaus and cold sea currents, its effects are less obvious than in other parts of the Arctic, at least for now.

This huge region has more than 100 species of mammals, 25 of them marine. Two species (Maritime Shrew and Labrador Collared Lemming) are endemics; a few others are either easier to see here than elsewhere (like Narwhal) or look particularly beautiful (like the white Arctic race of Gray Wolf).

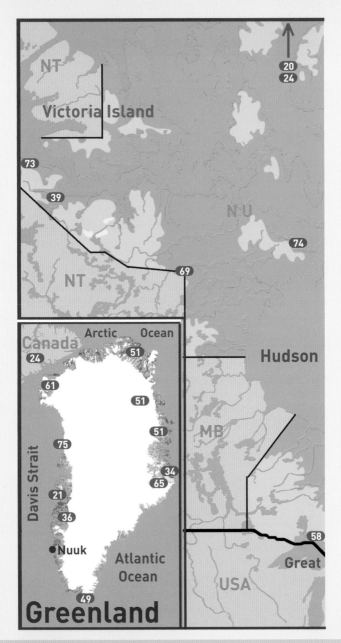

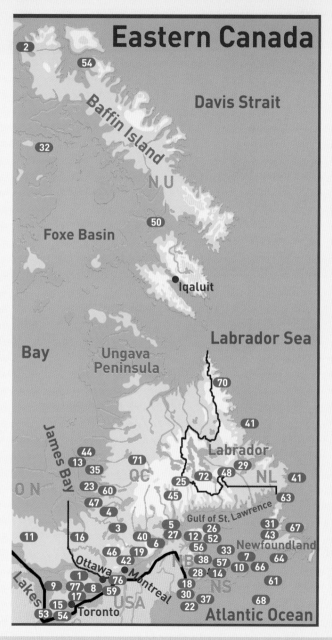

Eastern Canada

Davis Strait

Baffin Island

NU

Foxe Basin

• Iqaluit

Bay

Ungava Peninsula

Labrador Sea

Labrador

James Bay

QC

NL

ON

Gulf of St. Lawrence

Newfoundland

Ottawa

Montreal

NB

NS

Lakes

USA

Toronto

Atlantic Ocean

Eastern Canada, Nunavut, and Greenland

1. Algonquin PP (ON)
2. Arctic Bay (NU)
3. Ashuapmushuan FR (QC)
4. Assinica WR (QC)
5. Baie-Comeau (QC)
6. Baie-Sainte-Catherine (QC)
7. Bay St. Lawrence (NS)
8. Bonnechere Caves (ON)
9. Bruce Peninsula NP (ON)
10. Cape Breton Highlands NP (NS)
11. Chapleau (ON)
12. Chic-Choc Mts.; Gaspésie NP (QC)
13. Chisasibi (QC)
14. Confederation Bridge (NS/PE)
15. Crawford Lake CA (ON)
16. Aiguebelle NP (QC)
17. Darlington PP (ON)
18. Deer I. (NB)
19. Laurentides WR (QC)
20. Devon I. (NU)
21. Disko Bay (Greenland)
22. East Ferry (NS)
23. Eastmain Rd. (QC)
24. Ellesmere Island NP (NU)
25. Fermont (QC)
26. Forillon NP (QC)
27. Forrestville–Rimouski Ferry (QC)
28. Fundy NP (NB)
29. Goose Bay (NL)
30. Grand Manan I. (NB)
31. Gros Morne NP (NL)
32. Iglulik (Igloolik) (NU)
33. Iles de la Madeleine (Magdalen Islands) (QC)
34. Ittoqqortoormiit (Greenland)
35. James Bay Rd. (QC)
36. Kangerlussuaq (Greenland)
37. Kejimkujik NP (NS)
38. Kouchibouguac NP (NB)
39. Kugluktuk (NU)
40. La Baie Sainte-Marguerite (QC)
41. Labrador ferries (NL)
42. La Mauricie NP (QC)
43. Little Grand Lake Provincial Ecological Reserve (NL)
44. Longue Pointe (QC)
45. Manicouagan Impact Crater (QC)

Little Brown Myotis is the most common bat in caves of Eastern Canada.

46. Mastigouche WS (QC)
47. Matagami (QC)
48. Muskrat Falls (NL)
49. Nanortalik (Greenland)
50. Nettilling Lake (NU)
51. Northeast Greenland NP (Greenland)
52. Percè (QC)
53. Pinery PP (ON)
54. Pond Inlet/Mittimatalik (NU)
55. Port Burwell PP (ON)
56. Port-Daniel WS (NB)
57. Prince Edward Island NP (PE)
58. Pukaskwa NP (ON)
59. Rideau River PP (ON)
60. Route du Nord (QC)
61. Qaanaaq (Greenland)
62. Sable I. (NS)
63. St. Anthony (NL)
64. St. Pierre I.
65. Scoresbysund Fjords (Greenland)
66. Sydney Harbor (NS)
67. Terra Nova NP (NL)
68. The Gully MPA (NS)
69. Thelon WS (NT/NU)
70. Torngat Mts. NP (NL)
71. Trans-Taiga Rd. (QC)
72. Trans-Labrador Hwy. (NL)
73. Tuktut Nogait NP (NU/NW)
74. Ukkusiksalik NP (NU)
75. Upernavik (Greenland)
76. Voyageur PP (ON)
77. Warsaw Caves CA (ON)

The southern part of eastern Canada is covered by mixed and coniferous forests, with hardwood forests in the warmest parts and extensive tundralike bogs on many hilltops. Ontario is the most developed part of Canada, although even here the wilderness is never too far. In **Ottawa's** suburbs there are plenty of interesting mammals, including Fisher, Eastern Cottontail, black-morph Eastern Gray Squirrel, and occasionally Algonquin Wolf. **Voyageur Provincial Park** just off Autoroute 40 is a good place to see Northern Raccoon.

Probably the best mammal-watching destination in the province is **Algonquin Provincial Park**, where Fisher, Least Weasel, American Mink, Northern River Otter, Eastern Chipmunk, Woodchuck, Northern and Southern Flying Squirrels, Deer Mouse, Southern Red-backed Vole, and Woodland Jumping Mouse are common, and Moose is more reliable than anywhere in eastern Canada. Algonquin Wolf (locally called "Eastern Wolf") is also often seen, especially in winter and early spring in the areas just outside the eastern border of the park, where large numbers of White-tailed Deer concentrate, as well as along the main park road and near the main visitor center. The park sometimes places deer carcasses near the visitor center in winter months (particularly during spring school break) to attract wolves. Bird feeders near the visitor center are often visited by Fisher and

Fall color season is the best time for mammal watching in Eastern Canada; the weather is often better than in summer.

American Marten in December through early March. Southern Bog Lemming can be seen along Spruce Bog Boardwalk Trail. Check the wildlife sighting boards at the visitor center and the eastern entrance for recent sightings.

New Brunswick has two nice national parks, **Kouchibouguac** and **Fundy**, where a variety of forest mammals, from jumping mice and Southern Red-backed Vole to Northern River Otter and Moose, can be seen. In the latter park, you can observe Gray Seals coping with the world's highest tides; Maritime Shrew also occurs there. **Deer Island** near the U.S. border is famous for the world's second-largest whirlpool just off its southern tip; you can paddle a kayak into the center of the whirlpool and observe feeding Gray Seals and Harbor Porpoises up close. Whale-watching tours from **Grand Manan Island** (with **Sea Watch Tours**) are the best option for seeing Northern Right Whale, present only in late summer. Northern Minke, Humpback, and Fin Whales, as well as Atlantic White-sided Dolphin, Short-beaked Common Dolphin, and Harbor Porpoise, are also seen regularly. All of them are occasionally observed from the ferry to Grand Manan, and rarely from Deer Island ferries.

Nova Scotia also has great national parks. **Kejimkujik National Park** is good for forest species, such as Southern Red-backed Vole, while **Cape Breton Highlands National Park** is great for shrews, Coyote, Bobcat, Rock Vole, and Northern and Southern Bog Lemmings, as well as for great vistas. American Mink is common on rocky shores around the peninsula at low tide. Whale-watching tours from **East Ferry** in the south regularly encounter Northern Right, Fin, and Humpback Whales, while those from **Bay Saint Lawrence** are good for Humpback, Northern Minke, and Fin Whales; Long-finned Pilot Whale; and Atlantic White-sided Dolphin. Late summer is the best whale-watching season. White-beaked Dolphins are sometimes visible in summer from **Confederation Bridge** connecting Prince Edward Island and New Brunswick.

Remote and very difficult to access, **Sable Island** has the largest breeding colonies of Harbor and Gray Seals in the North Atlantic, while Northern Bottlenose Whale and Pygmy Sperm Whale occur offshore. The only land mammal on the island is feral Horse. Even more difficult to access is **The Gully Marine Protected Area**, a large submarine canyon east of Sable Island, where Northern Bottlenose Whale, Sowerby's Beaked Whale, and Sperm Whale regularly occur in summer.

Northern Right Whales can be seen in late summer off Nova Scotia.

The coastline becomes stunningly beautiful as you cross into Québec. The best stretch begins near **Percé**. Whale-watching and scuba diving tours are run from here in summer to **Bonaventure Island**, famous for its huge gannet colony. Blue, Northern Minke, and Humpback Whales and Atlantic White-sided Dolphin are commonly seen, while Gray and Harbor Seals are abundant above and below the surface. A bit farther north, **Forillon National Park** has the Gaspé race of Rock Shrew; numerous more common species, such as American Black Bear and Eastern Chipmunk; and some of the last surviving Eastern Cougars. Harbor Porpoise is often visible offshore. **Gaspésie National Park** in the Chic-Choc Mountains has more northern fauna, including abundant Moose, Snowshoe Hare, Canada Lynx (relatively common, especially following years of good Snowshoe Hare numbers), and Woodland Caribou on mountaintops (best seen on Mt. Albert). The park has excellent hiking trails but few roads, so it's better to conduct night drives on forest roads to the east of it.

The Gulf of St. Lawrence is a broad, cold, windy fjord with numerous islands. A number of ferries cross it; you can occasionally see whales from these ferries (particularly from **Forrestville–Rimouski** and **Trois Pistoles–Les Escoumins Ferries**), but it's better to take a whale-watching tour. Tours from **Baie-Sainte-Catherine** are good for Blue and Northern Minke Whales; those from **Tadoussac** for Humpback and Northern Minke Whales, Beluga, and Harbor Porpoise. It might also be possible to see whales from the shore: try **La Baie Sainte-Marguerite** and **Cap du Bon** for Beluga and Northern Minke Whale, and the latter also for Harbor Porpoise. Harp Seal is sometimes visible from shore around Tadoussac in spring. **Îles de la Madeleine** (the Magdalen Islands) are a group of low, sandy islands in the middle of the Gulf. Red Fox and Snowshoe Hare are common there, but the main reason for visiting the islands is the availability of helicopter tours to seal pupping areas on nearby ice floes. These tours are run by the Château Madelinot hotel for two to three weeks a year in early March. Harp Seal is the main attraction; Hooded Seal is rare, spends only a few days on ice, and is not always seen, so you have to tell the hotel owners and helicopter pilots in advance that you are looking for it. These trips require some complicated planning, as you need to come to the island for at least four days in case the weather gets too bad for flying, and Hooded Seal pupping season is really difficult to catch.

The most challenging driving adventure in eastern North America involves driving the road from **Baie-Comeau** in Québec to **Goose Bay** in Labrador, partially known as the **Trans-Labrador Highway**. It takes a few days of travel through endless boreal forest, where American Black Bear, Red Fox, Moose, Snowshoe Hare, Northern Porcupine, and occasionally Canada Lynx are seen. The road passes through the colossal, ring-shaped **Manicouagan Impact Crater**, where Least Weasel is common in lakeside bogs. Then it crosses a high pass on the Québec/Labrador line, where tundra slopes are inhabited by Labrador Collared Lemming and Eastern Heather Vole. The stretch near Muskrat Falls turnoff is the best place to look for Eastern Water Shrew; the falls area is good for Rock Vole, and the vicinity of Goose Bay for Least Weasel and Northern Bog Lemming. From Goose Bay you can either return to Baie-Comeau or take another marathon drive south along the coast, then a ferry to Newfoundland and another (very expensive) ferry to Nova Scotia.

In Goose Bay you can try whale watching by yourself by taking a ferry to coastal villages farther north. Humpback, Northern Minke, Fin, Northern Bottlenose, and Sowerby's Beaked Whales, as well as Beluga, are possible along the coast. You can also charter a flight to **Torngat Mountains National Park**, an outstandingly beautiful land

Northern Porcupines, ranging in color from blonde to almost black, are a common sight on the Trans-Labrador Highway.

of snow-clad mountains, glaciers, fjords, and tundra. Mammals there include American Black and Polar Bears, Gray Wolf, Red and Arctic Foxes, Caribou, a few species of voles, and Labrador Collared Lemming. Getting anywhere in the park requires very long hiking trips.

Another adventurous driving trip in northern Québec is to the eastern coast of James Bay. It is shorter than the drive to Goose Bay, but still requires some endurance, particularly in late fall and winter when the roads are covered with ice and driving is slow. Sometimes you see nothing interesting for hours or even days, but when you finally see wildlife, it can be really spectacular. Once an American Marten chased a Snowshoe Hare under the wheels of my car; I had to brake so hard that the car almost spun off the road.

Gray Wolves are traditionally hated and feared in much of the U.S., but the Canadian public is much more tolerant of them, and political controversies surrounding wolves are largely a thing of the past in Canada.

You can start the journey by visiting **La Mauricie National Park**, a good place to look for American Mink, Fisher, Northern River Otter, Deer Mouse, and, with some luck, Star-nosed Mole (try the area around the first bridge on Mekinac Trail). The next large protected area is **Ashuapmushuan Wildlife Reserve** (Réserve Faunique Ashuapmushuan), where Northern Flying Squirrel occurs and Dwarf Shrew can be found in roadside ditches. Past Chibougamau you enter the northern wilderness, so you should be self-sufficient. An unpaved road called Route du Nord leads into the huge **Assinica Wildlife Reserve**, where American Marten, Least Weasel, American Mink, Eastern Heather Vole, and Snowshoe Hare occur. The road eventually joins James Bay Road, which runs parallel to the bay shore but about 60 to 100 miles inland. It is connected to the bay by unpaved side roads, such as **Eastmain Road**, where Woodland Caribou can sometimes be seen in spring and fall, while Gray Wolf, Northern Bog Lemming, and other northern species occur year-round. The northernmost of these connecting roads runs from Radisson at the end of James Bay Road to **Longue Pointe**. Hilltops in this area are covered with dry tundra, where Labrador Collared Lemming, Snowshoe Hare, Ermine, and possibly Wolverine and Arctic Hare can be found. In winter, Arctic Fox and sometimes Polar Bear migrate to James Bay coast from the north, while Ringed and Bearded Seals can be seen on ice from Longue Pointe and other coastal settlements. Avoid driving the last few miles to the bay in November through early December, when lake effect causes almost nonstop blizzards there. You can return to civilization by driving James Bay Road to **Matagami**, where Red Fox can often be seen patrolling the streets at night. The last place worth visiting on this route is **Aiguebelle National Park** (Parc National d'Aiguebelle), where you have a good chance of seeing Moose, White-tailed Deer, Coyote, and Rock Vole (in canyons).

If you are truly adventurous, you can extend this driving trip by exploring the **Trans-Taiga Road**, which branches off James Bay Road and reaches the mining outpost of Caniapiscau 414 miles (666 km) to the east. This unpaved road is extremely rough and remote, and it's a good idea to carry a couple of extra spare tires, or, better, spare wheels, as well as at least 10 gallons of spare fuel. A 4x4 vehicle is recommended past Brisay (mile 362/km 582). Trans-Taiga Road doesn't reach true tundra, but there are tundralike patches on hilltops and numerous lowland bogs where Northern Bog Lemming is common. This road is one of the best places in North America to look for Wolverine. Sightings of Gray Wolf, Woodland Caribou, and (in winter) Arctic Fox are common.

Newfoundland is an interesting mammal-viewing destination. Huge pupping concentrations of Harp and Hooded Seals form on nearby ice floes, but you need to charter a helicopter to visit them. In June through early July, Long-finned Pilot, Humpback, Northern Minke, and sometimes Fin Whales congregate in **Bonne Bay** in **Gros Morne National Park** and can be watched from a kayak. Many rare species of whales occur offshore, but the numerous whale-watching operators go only a short distance from the coast, so the only species seen regularly are the same four, although Beluga and Killer Whale are also possible. White-beaked Dolphin is often seen in late summer and early fall during whale-watching tours from **St. Anthony**.

There are no endemic mammals on Newfoundland, but the local races of Red Fox and American Marten are very distinctive. **Little Grand Lake** area is the best for Marten, while Red Fox is common all over the island. Gros Morne National Park is a good place to see

Woodland Caribou (on mountaintops in summer and on the coast north of Berry Hill in winter), Moose (introduced), Snowshoe Hare, Arctic Hare (on Gros Morne Mountain), American Black Bear of the second-largest subspecies (try Stuckless Pond, Lomond, Snug Harbor, and Gros Morne Mountain Trails), and sometimes Canada Lynx. It is also the best place in North America to look for Masked Shrew. Harbor Seals haul out in summer in St. Paul's Inlet within the park; seal-viewing boat tours are available. American Mink, Moose, and Long-finned Pilot Whale are also common in **Terra Nova National Park**.

Nunavut is even more remote than Labrador and northern Québec. Flying there from Ottawa is ridiculously expensive, and since you can't see all wildlife in any single location, you have to spend even more on local flights and boat trips. But the landscapes are sublime, and wildlife-viewing opportunities fantastic. It might be cheaper to obtain a pilot's license and rent a small airplane for the trip (airplane rentals are cheaper than most people think because you pay for only the time the engine is running), or simply take a ship cruise.

The most popular destination in Nunavut is **Pond Inlet** (**Mittimatalik**) on Baffin Island. As in many local towns, Nearctic Collared and Brown Lemmings and Ermine can be seen around houses. Arctic Fox, Gray Wolf, Barren Ground Caribou, and Arctic Hare occur in the tundra. In June you can arrange a boat to the edge of ice floes to see Ringed Seal, Beluga, and Narwhal. Polar Bear, Walrus, Bearded Seal, and Bowhead Whale are also possible, while Harp Seal is common later in season. In **Igloolik** and **Arctic Bay**, Polar Bear, Walrus, Bearded Seal, Narwhal and Bowhead Whale are more common. Arctic Kingdom (arctickingdom.com) organizes Arctic Bay–based tours that include snorkeling with Narwhals and Belugas. **Ellesmere Island National Park**, accessible from Resolute (**Qausuittuq**), has beautiful Peary Caribou, Muskox, Arctic Wolf, Polar Bear, and Nearctic Collared Lemming. **Kugluktuk** in far western Nunavut is a good place to look for Grizzly Bear, Wolverine, Muskox, and Barren Ground Caribou.

One of the largest and least visited protected natural areas in North America is the extremely remote **Thelon Wildlife Sanctuary**, shared by Northwest Territories and Nunavut. It has large populations of many taiga and tundra mammals, including Grizzly Bear, Gray Wolf, Wolverine, Barren Ground Caribou (migrating in immense herds), Moose (often occurring on the open tundra in summer), Muskox, Arctic Squirrel, and Nearctic and Richardson's Collared Lemmings.

Pine Squirrel is usually the first mammal you see and hear in the northern woods.

The only practical way to visit the roadless expanses of the sanctuary is to fly to the town of Baker Lake (a.k.a. Qamani'tuaq), arrange a bush plane flight to the place where the Thelon River enters the sanctuary, and float by a canoe or a kayak back to Baker Lake (a distance of about 300 miles). Some people prefer to arrange an airlift from the place where the river leaves the sanctuary to avoid paddling across large, windy lakes between the sanctuary and the town, but this almost doubles the already high price. It is an arduous and risky expedition, but in terms of wildlife viewing it is probably the best canoe trip in the world.

Greenland is another part of the world where tourism potential is great, but high airfares make it almost impossible to get anywhere. Arctic Fox, Ermine, Barren Ground Caribou, Arctic Hare, and Nearctic Collared Lemming are common throughout the nonglaciated portions of the island, while Humpback, Bowhead, and Killer Whales, as well as Beluga and Narwhal, are often seen in summer in the surrounding waters. In the far south of the island, near **Nanortalik**, Sei Whale and White-beaked Dolphin also occur. **Qaanaaq** and **Disko Bay** are good places to see Narwhal in spring, Fin Whale in summer, and Beluga in the fall. Boat tours from **Upernavik** are a good way to see Polar Bear, Walrus, and Ringed and Bearded Seals; Harp Seal is also common in late summer. Muskox is easily seen around **Ittoqqortoormiit** and near **Kangerlussuaq Airport**. An immense labyrinth of beautiful fjords called **Scoresbysund** stretches from Ittoqqortoormiit area into **Northeast Greenland National Park**, one of the world's largest and most expensive to access. This area has Polar Bear, Walrus, Narwhal, Muskox, and Ringed, Bearded, and Hooded Seals.

Cruise ships regularly visit Greenland and Nunavut; their itineraries are somewhat optimized for viewing large mammals. In August, Polar Bear and Ringed, Bearded, Hooded, and Harp Seals are almost always encountered; Walrus, Narwhal, and Bowhead Whale are likely; Fin, Humpback, and Killer Whales, Beluga, and Atlantic White-sided and White-beaked Dolphins are also possible. A visit to a Muskox herd is included in some itineraries, and there is a chance of seeing Arctic Fox, Ermine, Arctic Hare, Brown Lemming, and Nearctic Collared Lemming during landings.

Western Canada

Western Canada has even more wilderness than eastern Canada and Alaska, but the best places are generally cheaper to access, thanks to the excellent roads crossing some high-elevation habitat, penetrating the tundra, and descending into coastal fjords. You still have to pay for boat or plane transport to explore remote coastal areas, and there is no cheap way to get to the Arctic Coast, but elsewhere all you need is a reliable car and high tolerance for weeks-long drives.

The climate of the region is highly variable; you can quickly get from the permanently wet coast to the arid interior valleys of British Columbia, then climb the rainy western slope of the Rockies and descend to the dry prairies beyond.

Western Canada has 150 mammalian species, 21 of them marine. Two or three (Vancouver Island Marmot and one or two collared lemmings, depending on which taxonomy you prefer) are endemic.

A good way to explore the region is to start from Calgary; drive north through Banff and Jasper National Parks; get onto the Alaska Highway; explore the Yukon and, if you have the whole summer, Alaska; then return via the Stewart-Cassiar Highway (see below), making short side trips to the coast. You can also start from Vancouver and do this trip in reverse. In other times of the year, sleeping

Hoary Marmot is a common sight in the Canadian Rockies.

in your car or a tent can be a survival test, and daylight time might approach zero, but all major roads are still open and the animals you see are more photogenic thanks to lush winter coats (of course, some species such as bats, bears, marmots, ground squirrels, and jumping mice hibernate in winter).

Let's start from **Vancouver**. Of all North American cities, it has the most scenic location thanks to the backdrop of snow-clad mountains. This part of British Columbia has some Pacific Coast mammals not found anywhere else in Canada, such as Olympic, Pacific, Marsh, and Trowbridge's Shrews; American Shrew Mole; Coast and Townsend's Moles; Western Red Bat; Western Spotted Skunk; Sewellel; Townsend's Chipmunk; and Creeping Vole. **Burns Bog** on the southern outskirts of the city is good for small mammals, including Vagrant and Trowbridge's Shrews, Townsend's Vole, and the very rare Olympic Shrew. Much of the bog is private property, but the best part is open to the public and protected in **Delta Nature Reserve**. **Manning Provincial Park** farther inland has Columbian Ground Squirrel and Creeping Vole.

Harbor Porpoise is often seen from ferries to gorgeous **Vancouver Island**, where endemic Vancouver Island Marmot can be seen

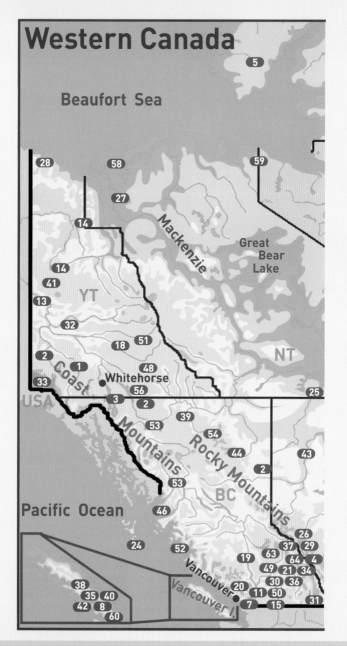

Western Canada

Beaufort Sea

5

Mackenzie

Great Bear Lake

28

58

27

14

59

14

41

13

32

18

51

YT

NT

2

1

48

Whitehorse

33

56

3

2

USA

25

53

39

54

44

43

2

53

BC

Pacific Ocean

46

24

52

26

37

29

19

63

64

4

49

21

34

30

36

20

11

50

7

15

31

Vancouver

Vancouver I.

38

35

40

42

8

60

Coast Mountains

Rocky Mountains

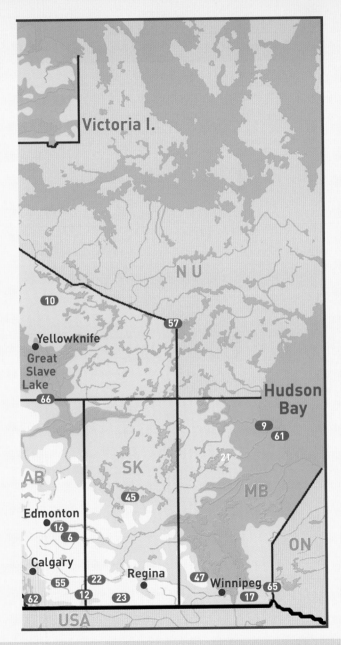

Victoria I.

N U

10

Yellowknife
Great
Slave
Lake

57

66

Hudson
Bay

9 **61**

AB

SK

21

MB

45

Edmonton
16
6

ON

Calgary
55

22

Regina

47 Winnipeg **65**

62

12

23

17

USA

Western Canada

1. Aishihik Lake (YT)
2. Alaska Hwy. (AB/BC/YT)
3. Atlin Rd. (YT)
4. Banff NP (AB)
5. Banks I. (YT)
6. Beaverhill Lake (AB)
7. Burns Bog (BC)
8. Cathedral Grove (BC)
9. Churchill (MB)
10. Contwoyto Lake ice rd. (NT/NU)
11. Coquihalla Canyon PP (BC)
12. Cypress Hills PP (AB/SK)
13. Dawson City (YT)
14. Dempster Hwy. (NT/YT)
15. E. C. Manning PP (BC)
16. Elk Island NP (AB)
17. Emerson (MB)
18. Faro (YT)
19. Fraser River (BC)
20. Garibaldi PP (BC)
21. Glacier NP (BC)
22. Great Sandhills (SK)
23. Grasslands NP (SK)
24. Gwaii Haanas NP (BC)
25. Hay River (NT)
26. Hwy. 40 (AB)
27. Inuvik (NT)
28. Ivvavik NP (YT)
29. Jasper NP (AB)
30. Kamloops (BC)
31. Kikomun Creek PP (BC)
32. Klondike Hwy. (YT)
33. Kluane NP (YT)
34. Kootenay NP (BC)
35. Mt. Modeste (BC)
36. Mt. Revelstoke NP (BC)
37. Mt. Robson PP (BC)
38. Mt. Washington (BC)
39. Muncho Lake PP (BC)
40. Nanaimo Harbor (BC)
41. Ogilvie Mts. (YT)
42. Pacific Rim NP (BC)
43. Peace River Valley (AB)
44. Pink Mt. (BC)
45. Prince Albert NP (SK)
46. Prince Rupert (BC)
47. Riding Mt. NP (MB)
48. Robert Campbell Hwy. (YT)
48. Seton Lake (BC)
50. Skagit Valley PP (BC)
51. South Canol Rd. (YT)
52. Spirit Bear Lodge (BC)
53. Stewart-Cassiar Hwy. (BC/YT)
54. Stone Mt. PP (BC)
55. Suffield NWA (AB)
56. Teslin Lake (YT)
57. Thelon WS (NT/NU)
58. Tuktoyaktuk (NT)
59. Tuktut Nogait NP (NT/NU)
60. Victoria (BC)
61. Wapusk NP (MB)
62. Waterton Lakes NP (AB)
63. Wells Gray PP (BC)
64. Yoho NP (BC)
65. Whiteshell PP (MB)
66. Wood Buffalo NP (AB/NT)

at **Mount Modeste** and around **Mount Washington** Alpine Resort. Keen's race of Long-eared Myotis can be found in colossal hollow Douglas firs in **Cathedral Grove**, while Keen's Mouse is common at the edges of forest clearcuts. Roosevelt Elk and Snowshoe Hare are common throughout the island. **Telegraph Cove** is a good place to see Harbor Seal, Harbor Porpoise, and sometimes Killer Whale. In winter and spring, Gray Whale is often visible offshore, while Steller's Sea Lions haul out in **Nanaimo Harbor**. Whale-watching tours from **Victoria** are good for Killer, Northern Minke, Gray, and Humpback Whales and Dall's Porpoise.

The **Kamloops** area in the interior is the arid heart of otherwise well-watered British Columbia; Yellow-bellied Marmot is common here in dry canyons. Very scenic Caribou Highway (Highway 97) goes straight north toward Prince George (you can make a short side trip to see Mountain Goat at **Seton Lake**). By the time you pass Prince George and turn onto Highway 16, you no longer pay attention to Moose, Red Fox, and Northern Porcupine on the roadsides, but the best is yet to come.

Highway 16 ends in Prince Rupert, the only convenient access point to much of the mainland British Columbia coast. From here you can take a ferry to **Haida Gwaii** (formerly Queen Charlotte Islands), where the largest subspecies of American Black Bear is easy to see, as are American Marten and coast-dwelling Northern River Otter; introduced Mule Deer is abnormally abundant here, and Keen's Mouse can be seen hunting for sand hoppers on rocky shores at night. You can also charter a boat to go south from Prince Rupert and try to see Kermode Bear (the white morph of American Black Bear) fishing in salmon rivers. **Spirit Bear Lodge** is said to be a good, albeit expensive, place to see Kermode, as well as conventional Black and Brown Bears, Gray Wolf, and the Sitka subspecies of Mule Deer.

Just 93 miles (150 km) before Prince Rupert (very close by Canadian standards), Highway 37 (**Stewart-Cassiar Highway**) branches off to the north. This is the best road in Canada for viewing bears; in early summer, you can see two or three American Black Bears per mile on some stretches as they feed on young grass on the shoulders. Grizzly Bear (mostly in spring and fall), American Marten, Gray Wolf, and Canada Lynx are also seen sometimes. In winter, herds of Woodland Caribou occur around mile 865 (km 1,392). Pristine wilderness stretches for hundreds of miles on both sides of the road; hotels are rare, and some have signs like "Put some money in the mailbox and take any room." A side road, short but dangerous because of frequent avalanches, provides access to the coast at Stewart (you have to cross the border into the tiny U.S. town of Hyder, Alaska, to see the ocean). This is the access point to Misty Fjords National Monument (see the next section on Alaska).

The Stewart-Cassiar Highway eventually crosses into Yukon Territory and joins the **Alaska Highway**. Look for Stone Sheep at mile 822 (km 1,323), for Rocky Mountain Elk around mile 948 (km 1,526), and for Moose, Woodland Caribou, Northern Flying Squirrel, and American Beaver everywhere. There are two more access roads to the coast, both entering Alaska via high passes where Dall Sheep is sometimes visible, Tundra Vole is common around tundra lakes, and both Collared Pika and Hoary Marmot can be seen on talus slopes. One of these roads leads to Skagway and another one to Haines, where you can rent a plane for a flight (short, outstandingly scenic, and relatively cheap) to Glacier Bay National Park (see next section) or get onto a ferry to other parts of Alaska or to Seattle. Look for Grizzly Bear around mile 113 (km 182) of Haines Road. Another side road from the Alaska Highway goes to **Aishihik Lake**, where Wood Bison, the largest land mammal in the Americas, can be seen. Both Haines Road and the Alaska Highway skirt the edges of **Kluane National Park**, a land of extensive glaciers and countless lakes. Long, difficult, but spectacular hikes into the park lead to places where Dall Sheep, Mountain Goat, and American Beaver are easy to observe.

Instead of driving straight into Alaska, you can turn north toward Dawson at Whitehorse. In summer this road (Klondike Highway) can also be driven all the way to Alaska, where it eventually rejoins the Alaska Highway. Look for Coyote and Woodchuck along the way. A side trip to **Faro** is the only easy way to see Fannin Sheep; Canada Lynx and Coyote are also common in that area, and Stone Sheep occur nearby along South Canol Road.

Just before Dawson, unpaved Highway 5 (**Dempster Highway**) branches off for a spectacular run to Inuvik in Northwest Territo-

American Pika in Waterton Lakes National Park, Alberta.

ries, 457 miles (736 km) away. In winter you can continue for another 125 miles (200 km) to Tuktoyaktuk, the only place in North America where you can drive straight to the Arctic Ocean (there are plans to build an all-season road there). In late spring and late fall you might not be able to get even to Inuvik, as there are gaps of a few weeks when neither ice crossings nor ferries across Peel and Mackenzie Rivers are operational. You cross the Arctic Circle along the way, so in late June you can enjoy 24-hour sunlight, and in late December continuous darkness. There are only three gas stations on that road, and traffic is much lighter than on somewhat similar Dalton Highway in Alaska, so you have to be self-sufficient and have some courage to travel this route, especially in winter. Herds of Woodland Caribou and a few Moose occur year-round along the Yukon part of the road, while Porcupine Caribou can sometimes be seen past the Mackenzie River crossing. In spring you can see Grizzly Bear and sometimes Gray Wolf and Wolverine feeding on roadkill. Collared Pika, Ogilvie Mountains Collared Lemming, and Hoary Marmot occur in the Ogilvie Mountains, and can be seen by walking a short distance from the road at miles 51 to 53 (km 82 to 85). (It is possible that Alaska Marmot is also present.) Once the road enters the lowland tundra, look for Tundra Shrew, Wolverine, Arctic Fox, Arctic Ground Squirrel, Singing and Tundra Voles, and Common Muskrat. Tundra and Barren Ground Shrews, Snowshoe Hare, Common Muskrat, Brown Lemming, and Nearctic Collared Lemming are common around Inuvik. Beluga can often be seen in the Mackenzie River near the town in late summer. You can find air transport to **Banks Island** to see Muskox, as well as to **Ivvavik** and **Tuktut Nogait National Parks**; or (much cheaper) to Tuktoyaktuk, where you can arrange a tour to see Porcupine Caribou, Polar Bear, Ringed and Bearded Seals, Beluga, and (in summer and early fall) Bowhead Whale.

If you are still capable of driving by the time you return to the Alaska Highway, you can head southeast, to pass through some rather boring forest landscapes, interrupted by three beautiful mountain ranges. **Muncho Lake Provincial Park** and **Stone Mountain Provincial Park** are the most accessible places to see Stone Sheep; look also for Woodchuck, Bushy-tailed Woodrat, and Woodland Jumping Mouse. If you have a high-clearance vehicle (preferably rental or fully insured), you can drive to the summit of **Pink Mountain**, where Grizzly Bear, Woodland Caribou, Hoary Marmot, and Northern Bog Lemming occur in summer.

As you enter Alberta, you can make yet another long, exhausting, but exhilarating side trip into the northern wilderness. It takes two full days of nonstop driving to get to huge **Wood Buffalo National Park**, shared by Alberta and Northwest Territories. The best time for the trip is late June or early July, when there is almost 24-hour daylight and you can often see nocturnal animals, such as Fisher, Canada Lynx, Northern Flying Squirrel, and American Beaver, that have no choice but to be active during the day. Gray Wolf is more often seen in late winter when there is a lot of snow and it prefers to use roads. In the park you can see impressive herds of Wood Bison. Arctic Shrew, Snowshoe Hare, Yellow-cheeked Vole, and other typical mammals of the northern woods are common. The park is also famous for the world's largest beaver dams. In winter you can use the ice road from Fort McMurray to access the southeastern corner of the park, where fascinating interactions between Gray Wolf packs and Wood Bison herds can often be observed.

The world's longest ice road is built every winter along a chain of frozen lakes stretching for 373 miles (600 km) from Tibbitt Lake near Yellowknife to **Contwoyto Lake** in the tundra of Nunavut. It is operational in only February and March, and can be hazardous to drive, but Wolverine is reportedly seen sometimes near roadkill or gut piles left by hunters along that road. Barren Ground Caribou and Gray

Wood Bison is the largest terrestrial animal in the Americas.

Thanks to conservation efforts that required a lot of self-sacrifice from the Native people, Muskox is again common in parts of the Arctic.

Wolf are also frequently spotted there. Northern Red-backed Vole is common around Yellowknife in summer.

One of the most beautiful parts of the world is the chain of protected areas covering much of the Canadian Rockies, from Willmore Wilderness Park in the north to Waterton Lakes National Park on the U.S. border. The parks on the western (British Columbia) side of the mountains, such as **Mount Robson Provincial Park**, **Yoho**, **Kootenay**, **Glacier**, and **Mount Revelstoke National Parks**, are lusher, with glaciers descending into tall rainforests. Southern Red-backed Vole and marsh-loving mammals such as Cordilleran Water Shrew and American Mink are easier to see there, but the weather is often really bad. Parks on the eastern (Alberta) side are a bit drier, but there are still plenty of glaciers, some with easy road access, and outstanding abundance of wildlife, particularly in **Jasper** and **Banff National Parks**. In winter, look for Rocky Mountain Bighorn Sheep along Highway 16 a mile or two northeast of Jasper, for Woodland Caribou along the road to Wapiti Campground, and for Gray Wolf and Cougar anywhere around Jasper. As you follow Icefields Parkway south toward Banff, look for more Caribou and Bighorn Sheep, Rocky Mountain Elk (some graze in the town of Banff), Mule Deer, Mountain Goat (at roadside salt licks), American Pika (look at mile 56 [km 90]), Long-tailed Vole, and Deer Mouse. Both Eastern and

Western Heather Voles occur in this area. In summer they are joined by Least Chipmunk (try mile 61.5 [km 99]) and Golden-mantled and Columbian Ground Squirrels. **Waterton Lakes National Park** is said to be a good place to look for Cougar and Wolverine in winter, and for Red-tailed Chipmunk in summer, although I never managed to see them here.

Compared to the splendid mountains, eastern Alberta, Saskatchewan, and southern Manitoba with their plowed prairies and monotonous forests seem boring. But here, too, are some places well worth visiting. **Elk Island National Park** in Alberta has easy-to-see Manitoba Elk, Northern Pocket Gopher, and Richardson's Ground Squirrel (there are also fenced herds of Plains and Wood Bison). **Grasslands National Park** in Saskatchewan has the best variety of prairie animals in Canada; look for Prairie Shrew, reintroduced Black-footed Ferret, Plains Bison, Pronghorn, White-tailed Jackrabbit, Franklin's and Thirteen-lined Ground Squirrels, Black-tailed Prairie Dog, Olive-backed Pocket Mouse, Northern Pocket Gopher, Northern Grasshopper Mouse, and Sagebrush Vole. **Prince Albert National Park**, also in Saskatchewan, has many forest mammals, such as Gray Wolf,

Pronghorn is one of many prairie species present in Grasslands National Park in Saskatchewan.

Polar Bears of Hudson Bay spend their six-month-long summer break in the coastal tundra, fasting and entertaining tourists. But their survival is now highly problematic as the bay freezes over later in the fall, so that the bears have to fast longer.

Canada Lynx, Fisher, Ermine, Manitoba Elk, Woodland Caribou, and Southern Red-backed Vole, as well as a herd of Plains Bison in the southwestern part of the park near Amyot Lake. **Riding Mountain National Park** in Manitoba is a good place to see Pygmy Shrew, Moose, Manitoba Elk, Snowshoe Hare, Woodchuck, Southern Red-backed Vole, Eastern Heather Vole, Northern Bog Lemming, and Northern Porcupine.

Last but not least, **Churchill** on the coast of Hudson Bay, accessible by train or relatively affordable flights, is the cheapest place in the world to see Polar Bear (October through November). Most people take an ATV tour, but if you are on a really tight budget you can drive around yourself; sooner or later you'll find a bear or two. A rocky peninsula in the northern part of the town has Arctic and Red Foxes, Arctic Hare, and Richardson's Collared Lemming; forested inland uplands are inhabited by Snowshoe Hare, Deer Mouse, and Yellow-cheeked Vole; the airport area is good for Northern Bog Lemming (look also for lemmings crossing in front of the vehicle during bear tours); Boreal Water Shrew lives in wetlands. In winter, Ringed Seal is common just offshore. In summer, Beluga-watching tours are popular, and Arctic Ground Squirrel occurs around town.

Alaska

High mountains, rugged coastlines, and exceptional climate diversity make Alaska the most scenic part of the Arctic. They also make travel a bit difficult. You run out of roads to drive in about three weeks; after that you are limited to hiking trips, expensive ferries, and local flights (unless, of course, you are a local resident with a private airplane or a team of sled dogs). It is a good idea to start planning your trip at least

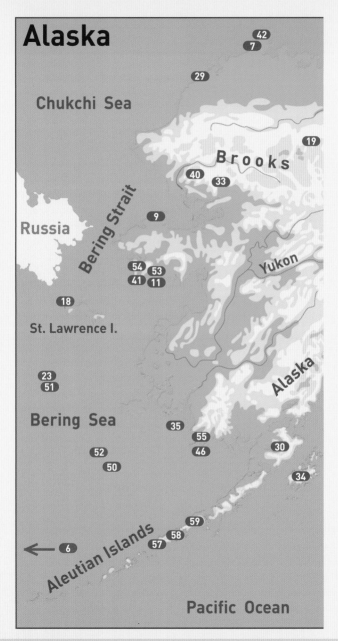

Alaska

Chukchi Sea

Russia

Bering Strait

Brooks

Yukon

Alaska

St. Lawrence I.

Bering Sea

Aleutian Islands

Pacific Ocean

42
7
29
19
40
33
9
54
53
41
11
18
23
51
35
55
46
30
34
52
50
59
57
58
6

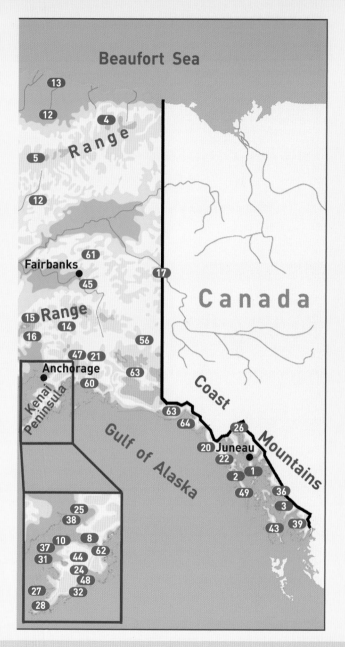

Beaufort Sea

13

12

4

Range

5

12

61
Fairbanks
45

15 Range
16 14

47 21
Anchorage
60

56

17

Canada

63

Kenai
Peninsula

Gulf of Alaska

63
64

Coast

26

20 Juneau
22

2 1

49

36

3

43 39

Mountains

25
38

10 8

37
31 44 62
24
27 48
28 32

Alaska

1. Admiralty I. NMT
2. Alexander Archipelago
3. Anan Wildlife Observatory
4. Arctic NWR
5. Atigun Pass
6. Attu I.
7. Barrow
8. Beluga Pt.
9. Bering Land Bridge NPR
10. Captain Cook NRA
11. Council Hwy.
12. Dalton Hwy.
13. Deadhorse
14. Denali Hwy.
15. Denali NP
16. Denali SP
17. Eagle
18. Gambell
19. Gates of the Arctic NP
20. Glacier Bay NP
21. Glenn Hwy.
22. Gustavus
23. Hall I.
24. Harding Icefield Trail
25. Hatcher Pass
26. Haynes
27. Homer
28. Kachemak Bay
29. Kasegaluk Lagoon
30. Katmai NP
31. Kenai
32. Kenai Fjords NP
33. Kobuk Valley NP
34. Kodiak I.
35. Kuskokwim Bay
36. LeConte Bay
37. Marathon Rd.
38. Matanuska-Susitna Valley
39. Misty Fjords NMT
40. Noatak NPR
41. Nome
42. Point Barrow
43. Prince of Wales I.
44. Resurrection Pass Trail
45. Richardson Hwy.
46. Round I.
47. Sheep Mt.
48. Seward
49. Sitka
50. St. George I.
51. St. Matthew I.
52. St. Paul I.
53. Taylor Hwy.
54. Teller Hwy.
55. Togiak NWR
56. Tok
57. Umnak I.
58. Unalaska I.
59. Unimak I.
60. Valdez
61. White Mts. NRA
62. Whittier
63. Wrangell-St. Elias NP
64. Yakutat

a year in advance, and watch for good deals on air tickets to remote locations. Always leave a couple of days between flights, since weather is often very bad. Some coastal villages in the south and on the Aleutian Islands have only five to ten sunny days per year; on the northern and western coasts snow might fall at any time. The central area has dry climate with moderately hot summers and the coldest winters on the continent. It also has the best fall colors anywhere in the Arctic.

Alaska has about 80 species of mammals (a bit more if you count island forms of uncertain status), 23 of them marine. At least one species (Alaska Marmot) is currently thought to be endemic, and three species are shared with only Asian countries.

Plains bison has been introduced to Alaska, where Wood Bison used to occur. It was a big mistake, since now Wood Bison can't be reintroduced there because of the danger of hybridization.

The southeastern Panhandle of Alaska is an insanely complex and breathtakingly spectacular labyrinth of fjords, islands, glaciers, and rugged mountains. There are few roads, except for three access roads from Canada (see previous section), short stretches radiating from towns, and a network of roads on **Prince of Wales Island** (a good place to look for Gray Wolf, American Black Bear, and the Sitka subspecies of Mule Deer). Most visitors see the area from cruise ships (including the so-called Alaska Ferries); accommodation and transportation can be outlandishly expensive, as many hotels, lodges, and rental companies mostly cater to trophy hunters and fishermen who tend to be little concerned with costs. Fortunately, there's plenty of space for camping, fjords are great for kayaking, and salmon can be bartered from local fishing camps. Extensive clearcut logging is rapidly destroying many local ecosystems, including huge swaths of the Tongass National Forest, so it's better not to delay your visit.

Anan Wildlife Observatory near Wrangell is an easily accessible place to watch American Black and Kodiak Bears fishing in late summer. **Sitka** is a convenient base for short trips in search of genetically distinct Brown Bears that inhabit islands in the Alexander

Archipelago, especially Baranof and Chichagof Islands. **Misty Fjords National Monument** in extreme southern Alaska, accessible from Hyder or Ketchikan, has Northern Elephant Seal, Steller's Sea Lion, Harbor Seal, Humpback Whale, and Dall's Porpoise in summer. American Black and Kodiak Bears are often visible from the water and from a viewing platform in Hyder. Mountain Goat is also common, and Cougar has been recorded.

Gustavus, accessible by short flights from Juneau, Skagway, and Haines, is the starting point for hikes and boat trips into gorgeous **Glacier Bay National Park**. This is a great place to see Harbor Seals near calving glaciers. American Black and Kodiak Bears are often seen along the shoreline as they feed on marine invertebrates and carrion; they are occasionally joined by Wolverine. Mountain Goat, Humpback Whale, Killer Whale, and Sea Otter are seen on most trips. The part of the park adjacent to Gustavus has American Marten and Glacier Bay Water Shrew along streams in mossy forests; Masked Shrew, Ermine, and the Sitka subspecies of Mule Deer around meadows; and Moose and Northern Red-backed Vole almost everywhere. In the town itself, look for American Mink, Northern River Otter, Hoary Marmot, and (in abandoned houses) Keen's Mouse.

Harbor Seal is often found near calving glaciers.

Yakutat is a little-visited town where you can arrange an expensive boat trip to see calving glaciers and the immense ice fields of Wrangell-St. Elias National Park. Harbor Seal and Sea Otter are usually present in Yakutat Bay, while American Black and Kodiak Bears and Northern River Otter are sometimes visible onshore. Driving at dusk along the single road out of town is a reliable way to see Moose and American Black Bear. If you are lucky, you can also spot Gray Wolf, Canada Lynx, and the silvery blue Glacier Bear, the most visually stunning land mammal in North America. Kodiak Bear can often be seen on town streets at night. American Marten, Snowshoe Hare, and Northern Red-backed Vole are common in the forest, while Tundra and Long-tailed Voles occur in coastal meadows.

Unless you arrive by a cruise ship or drive from Canada, your most likely starting point for Alaska travel is **Anchorage**. There is some nice mammal watching in and around town. At night, look for Moose and Northern Porcupine in Kinkaid Park near the airport. As you drive toward Kenai Peninsula, look for Common Muskrat at Potter Marsh, and for Beluga off Beluga Point (all in summer). In winter, Dall Sheep often graze near the road at Turnagain Arm.

Kenai Peninsula, a short drive southeast from Anchorage, is the only part of Alaska with a good road network. Moose, American Black Bear, and sometimes Gray Wolf can be seen from those roads and hiking trails. The best places to see Moose are Sterling Highway between Soldotna and Homer, Swan Lake Road, and the road to Captain Cook State Recreation Area, but you have to drive them before 6:00 a.m. Swan Lake Road is also good for American Beaver and Common Muskrat. Woodland Caribou can often be seen along Marathon Road in winter, and along Bridge Access Road through Kenai River Flats near the town of Kenai in August through September. Dall Sheep and Mountain Goat are often visible from Sterling Highway between Quartz Creek Campground and Cooper Landing. Pygmy Shrew, Northern Red-backed Vole, and Northern Bog Lemming are common in bogs and upland tundras. **Resurrection Pass Trail**, a 40-mile-long hiking trail between the town of Hope and mile 53 (km 85) of Seward Highway, provides a good chance to see Kodiak and American Black Bears, Woodland Caribou, and sometimes Gray Wolf and Wolverine. If you are in good shape you can walk the **Harding Icefield Trail** near Seward in one day; it reaches high-elevation habitat where American Black Bear, Mountain Goat, and Hoary Marmot are usually seen, especially in the morning. If you make it to the

Adult male Caribou with gorgeous antlers can often be seen in summer in Deadhorse, Alaska.

end of the trail, scan the ice field for Wolverine; it has been seen there a few times.

Inexpensive boat trips from Seward to **Kenai Fjords National Park** in summer provide an excellent opportunity to see Sea Otter, Harbor Seal, Steller's Sea Lion, Harbor and Dall's Porpoises, and Humpback, Northern Minke, and Killer Whales. Long-finned Pilot Whale and Gray Whale are also seen sometimes. Sea Otter and Steller's Sea Lion are often present around the jetty at the small boat harbor in Seward. Steller's Sea Lion, Harbor Seal, and Sea Otter can also be seen on short boat trips from Homer, and the latter are often swimming just offshore near the end of Homer Spit. Mountain Goat can also be seen on those trips if Sadie Cove is included in the itinerary. In May, Belugas are often close to shore in the town of Kenai and in Captain Cook State Recreation Area (best viewing on incoming tide).

Kodiak Bear is easier to see in more remote places farther south and west. Expensive bear-viewing tours depart from Anchorage, Kenai, and Homer and go to various salmon rivers on the western side of Cook Inlet, as well as to **Katmai National Park** and McNeil River State Game Sanctuary (which is terribly overcrowded by photographers). Cheaper tours are available on **Kodiak Island**, accessible by ferry from Homer.

The only place where you can reliably see Walrus onshore in Alaska without spending tens of thousands of dollars is **Round Island**, where a large rookery exists year-round. Overnight tours to the island depart from Dillingham, connected to Anchorage by plane. This is not a cheap excursion; reservations should be made months in advance.

Remote and sparsely populated, the **Aleutian Islands** are accessible by ferry from Anchorage. Unfortunately, the ferry schedule makes it necessary to make stopovers either very short or very long. As on other Alaska ferry routes, Sea Otter, Harbor Seal, Steller's Sea Lion, Dall's and Harbor Porpoises, and Northern Minke, Humpback, and Killer Whales are often seen. In addition, Aleutian waters have lots of Gray Whales in summer, and there is a chance of seeing Walrus, Baird's Beaked Whale, Fin Whale, Sei Whale, and (very rarely) North Pacific Right Whale. Stejneger's Beaked Whale occurs around the western part of the island chain, but ferries don't go there; you'd have to fly to **Attu Island** and go out with a fishing boat to look for it. Unfortunately, populations of many marine mammals in the Aleutians have recently crashed, most likely because of global warming. Land mammals (including Kodiak Bear) are mostly limited to the easternmost island of **Unimak**, although introduced Red Fox and

Some remote Alaskan roads get so little traffic that Red Foxes build their dens in road shoulders.

Caribou are common on many others. Tundra Vole and Unalaska Collared Lemming occur on **Umnak** and **Unalaska Islands**, the latter accessible by ferry.

Very few people travel to the Aleutians, but a lot of tourists visit central Alaska, where the weather is better and a neat circular driving tour via Fairbanks, Tok, and Glennallen is possible. If you have enough endurance you can drive this loop in a weekend, but it is better to devote at least two weeks to it, and make side trips up Dalton Highway (described below), into **Wrangell-St. Elias National Park** (it has lots of American Black and Brown Bears, Gray Wolf, Canada Lynx, Moose, Caribou, Singing Vole, and particularly Dall Sheep, but you have to hike above the timberline to see them in summer), and down to **Valdez** (the place to see Sea Otter, Harbor Seal, and Steller's Sea Lion if you are not visiting other coastal areas). Collared Pika can be found at **Hatcher Pass** above Wasilla, and Dall Sheep are always present on **Sheep Mountain** (miles 107 to 123 [km 172 to 198] on Glenn Highway).

The best place for viewing mammals in the Alaskan interior (and one of the best in the world) is **Denali National Park** and surrounding areas. To make the most of Denali, try to reserve a spot in Wonder Lake Campground; that will allow you to drive the entire length of the park road in your own car. Otherwise you are limited to buses, which can be recommended to only novice mammal watchers. Bus seats also have to be reserved in advance, especially on the first bus of the day, which is slightly better for wildlife viewing. The open tundra in the western part of the park is prime hiking country; you can see Montane and Eurasian Least Shrews, Porcupine Caribou, Grizzly Bear, Dall Sheep (mostly at Polychrome Pass), and sometimes Gray Wolf and Canada Lynx. Wolverine is seen there a few times per year. The eastern part is forested and has no driving restrictions; it is even easy to visit in winter (the western part can be accessed in winter only by skiing or by taking a dog sled tour). It is a good place to see Moose, Snowshoe Hare, Pine Squirrel, and Northern Porcupine. Tame American Beavers can be observed at dusk along Horseshoe Lake Trail. Savage River separates the two parts. A short hike up Savage River Loop Trail is a sure way to see Arctic Ground Squirrel and Hoary Marmot; Collared Pika is also common there in some years. Nearby **Denali State Park** has the same forest-associated species of mammals. If you missed Caribou and Northern Porcupine in Denali National Park, try driving **Denali Highway**, where Porcupines often

Arctic Ground Squirrel is delightfully easy to photograph in Denali National Park.

walk the shoulders and large herds of Caribou can sometimes be seen in the fall. Gray Wolf is said to be seen there frequently, and Wolverine occurs sometimes.

Dalton Highway is the most interesting driving route in Alaska. Websites devoted to this road often look like they try to scare people away, but in reality the road is easy to drive (you can safely speed up to 100 mph [160 kph] on some sections, although, of course, it's illegal) and reasonably safe (just pull off when a big truck is passing in the opposite direction, or you'll risk a cracked windshield). The only problem is that many rental companies specifically prohibit you from driving this road in their cars, so you might get in trouble if you have a breakdown or an accident. The first part of the highway passes through typical boreal forest where you don't see much wildlife, except for an occasional Moose, Red Fox, or Snowshoe Hare. If you walk into muskegs, you might be able to find Yellow-cheeked and Tundra Voles. At Atigun Pass, Dall Sheep and (in summer) Arctic Ground Squirrel are usually present. Beyond the pass it's mostly open tundra, where Masked Shrew, Wolverine, Arctic Fox, Gray Wolf,

Grizzly Bear, Porcupine Caribou, and Singing Vole are sometimes seen, although nothing is guaranteed. The steep mountain east of the highway at mile 259 (km 417) is the only easily accessible place to see Alaska Marmot. As you approach Deadhorse, look for Muskox east of the road, as well as for Brown Lemming and Nearctic Collared Lemming. In **Deadhorse** itself, Caribou occur in summer and Grizzly Bears visit garbage dumps in the fall. Unfortunately, you cannot drive to the Arctic Coast; the only way to see the water is to take a bus tour to Prudhoe Bay a few miles away. These tours are overpriced and have to be booked at least 48 hours in advance (thanks to a background check requirement). You are not going to see any new wildlife on those tours; Polar Bears sometimes wander into the town in winter and spring, but the tour buses avoid the places where they've been spotted. Dalton Highway skirts **Gates of the Arctic National Park** and **Arctic National Wildlife Refuge**. They have similar wildlife, but predators are more common and Polar Bear occurs on the coast. Visiting them requires chartering an airplane, except for the parts adjacent to Dalton Highway, which you can hike into.

To really see the Arctic Coast, you have to fly to **Barrow**, the northernmost town in the U.S. It is surrounded by very open Arctic tundra where larger mammals have been mostly hunted out, but Barren Ground and Tundra Shrews, Arctic Ground Squirrel, Northern Red-backed Vole, Beringian Brown Lemming, and Nearctic Collared Lemming still occur. Bowhead Whales pass just offshore in May, and Gray Whales in June. Native hunters can take you on a whaling expedition. Parts of whale carcasses are dumped onto a sandy spit called Cape Barrow east of town, which is popular with Polar Bears (they arrive in winter and sometimes linger until early summer). I recommend taking a tour to the spit, rather than walking, if there are bears present. As long as there are ice floes (approximately from October to early June, although this is changing now), you can often see Spotted, Ringed, and Bearded Seals, and sometimes Walrus from shore.

Nome is located on the treeless coast of the Bering Sea not far from the Bering Strait. As usual in Alaska backcountry, hotels and car rentals are ridiculously expensive, but if you talk to locals arriving with you on the plane or waiting at the airport, you should be able to find cheaper options. Three roads radiate from town. **Teller Highway** goes through coastal tundra and is good for Tundra Shrew, Red Fox, and Brown Lemming of the Beringian race. **Taylor Highway** (Kougarok Road) crosses slightly drier inland tundra and is better for the Alaska race of Arctic Hare, Red Fox, and Nelson's race of Nearctic Collared Lemming. **Council Highway** follows the coast, where Harbor and Spotted Seals congregate near river mouths in summer (Ringed Seal and Minke Whale are also seen sometimes). Singing Vole is abundant in some years around the "Trains to Nowhere" exhibit about 30 miles from Nome. From there the road turns inland and enters the forest, where Snowshoe Hare is very common in most years. A few herds of reintroduced Muskox occur in the hills around Nome (ask locals for recent locations, or start with Anvil Mountain and look around). American Beaver inhabits roadside ponds near Nome. In winter, Ringed, Spotted, and Bearded Seals can sometimes be seen on ice floes. In March through April it might be possible to charter a boat or a plane to go looking for beautiful and mysterious Ribbon Seal far offshore.

Nome isn't that remote by Alaskan standards, but the islands of the Bering Sea are. The Pribilof Islands have the world's largest Northern Fur Seal rookeries (over half a million strong), as well as blue-morph Arctic Fox. Both are easy to see on **St. Paul Island**. Unfortunately, the foxes are at their most beautiful in winter, while the fur seals are present only in summer, and the flights to the island aren't cheap. A distinctive race of Masked Shrew (often considered a separate species) occurs only on St. Paul (look in marshes and coastal tundra), while St. George has a distinctive race of Brown Lemming. Uninhabited

Northern Fur Seal babies can be easily observed on Pribilof Islands.

St. Matthew and **Hall Islands** also have blue Arctic Fox, as well as a distinctive race of Singing Vole, known as Insular Vole, and Walrus rookeries in summer. They are accessible only by private boat or small plane (there is a landing strip on St. Matthew). Zugunruhe Birding Tours (zbirdtours.com) is planning to run annual tours for advanced birders that will include St. Paul and St. Matthew Islands; these trips will also be a good chance to see rare marine mammals (possibly even Ribbon Seal). **St. Lawrence Island** (accessible by regular flights to Gambell) has a strikingly colored island race of Masked Shrew, also sometimes considered a separate species, as well as Tundra Vole and Northern Red-backed Vole. The waters surrounding all five islands are prime feeding grounds of Steller's Sea Lion and Gray Whale in summer, while Walrus and Spotted Seal are common year-round. Polar Bears often visit St. Lawrence Island when the sea is frozen.

Hawaii and Bermuda

Usually thought of as a tropical paradise, Hawaii is actually a sad place of environmental carnage. On the coast you can sometimes drive around for hours and never see a native plant. Forests are overrun by introduced ants, rats, and pigs. The majority of native species are extinct or endangered. It is a very comfortable country to travel around, with great climate and splendid vistas, but for a mammal watcher the best reason to visit is a chance to see some marine mammals that are easier to find here than elsewhere in North America.

The only native mammals in Hawaii are Hoary Bat, Hawaiian Monk Seal, and more than 20 species of whales and dolphins (of which only a few are seen regularly). Dog, Feral Pig, and Polynesian Rat were introduced by the Polynesians. Europeans brought numerous other species, of which Small Indian Mongoose, Black Rat, Norway Rat, and House Mouse are the most ubiquitous.

The island of Hawaii, commonly known as the Big Island, is the most wild and diverse. Hoary Bat can often be seen at dusk flying around Volcano House in **Hawaii Volcanoes National Park**. Feral Pig, Black Rat, and Polynesian Rat are common in and around the park. Feral Domestic Goat occurs on arid slopes. On the **Kona coast**, Small Indian Mongoose is frequently seen, and Humpback Whale is abundant just offshore from January to early April. Watch for breaching whales near Kailua-Kona Airport, or even from your plane just before landing.

Hawaii and Bermuda

1. Ahupua'a o' Kokena SP (HI)
2. Cross Seamount
3. Haleakala NP (HI)
4. Hawaii Volcanoes NP (HI)
5. Kailua-Kona (HI)
6. Kilauea Pt. NWR (HI)
7. Koke`e SP (HI)
8. Kona Coast (HI)
9. Lehua I. (HI)
10. Laysan I. (HI)
11. Makahuena Pt. (HI)
12. Mauna Kea (HI)
13. Midway Atoll NWR (HI)
14. Molokai I. (HI)
15. Na Pali Coast SP (HI)
16. Niihau I. (HI)
17. Paget Marsh NR (Bermuda)
18. Palmyra Atoll
19. Papahānaumokuākea Marine NWR (HI)
20. Poipu Beach (HI)
21. Puu Oo Crater (HI)
22. South Coast (Bermuda)
23. Waianae Coast (HI)
24. Waimea Canyon SP (HI)

Humpback Whale is the only whale species reliably seen around Hawaii in winter.

On Maui, **Haleakala National Park** has Hoary Bat and Small Indian Mongoose, but they are more difficult to find here than on the Big Island. Sugar cane fields along the coast have large populations of Black, Polynesian, and particularly Norway Rats; in some years they are joined by House Mouse in huge numbers.

Kauai, the most scenic island, is mongoose-free. Alakai Swamp Trail in **Koke`e State Park** is the best place to see Black and Polynesian Rats, as well as lots of native plants and animals. Hawaiian Monk Seals haul out a few times a week at **Poipu Beach**, at Ke'e and other beaches of **Na Pali Coast**, and in **Kilauea Point NWR**.

Humpback whale-watching tours are available in season (usually December through March) from numerous operators in all major cities; ask if the boat has a hydrophone (so that you can hear underwater sounds). More expensive dolphin-viewing tours are run from Kona, Lanai, Honolulu, and Lihue. Many tours provide an opportunity to snorkel with Spinner Dolphins, and offer a free second tour if dolphins are not found. Other commonly seen species include Common Bottlenose, Pantropical Spotted, and Rough-toothed Dolphins (the latter particularly common off Kauai). Short-finned Pilot, False Killer, Pygmy Killer, and Melon-headed Whales are seen a few times a year. Long-beaked Common, Striped, and Fraser's Dolphins; Cuvier's and Blainville's Beaked Whales; Sperm Whale; and Killer Whale have also been recorded. The best way to look for those rarities is to take the **Niihau Island** tour from Kauai with Port Allen–based Holo Holo Charters or a scuba diving operator. Hawaiian Monk Seals are seen during 70 to 80 percent of these tours, and sometimes approach snorkelers and divers at **Lehua Island**.

Less reliable alternatives to dolphin-viewing trips are watching the sea from **Kilauea Point NWR** or taking a morning ferry from Maui to Oahu, Lanai, or Molokai (afternoon rides can be very rough and spotting cetaceans becomes difficult).

The Northwestern Hawaiian Islands, a chain of small uninhabited islands stretching for over a thousand miles across the Pacific Ocean, have lots of Hawaiian Monk Seals. The surrounding waters have almost all whale and dolphin species of the tropical Pacific. However, these waters are protected as **Papahānaumokuākea Marine National Monument**, and visitation is extremely restricted. The only island not off-limits to the public is **Midway**, which has an airfield and permanent staff, but at the moment the only way to get there that is not outlandishly expensive is by volunteering for the annual seabird count.

The **Bermuda Islands** have no native land mammals, except for vagrants. Introduced Norway and Black Rats and House Mouse are common. In September through November, migrating North American bats are sometimes found here; the most often recorded ones are tree-roosting Seminole, Eastern Red, and particularly Hoary Bats (look in native Bermuda cedar trees in **Paget Marsh Nature Reserve**). In March through May, migrating Humpback Whales congregate around the islands. They are often visible from the south shore, and numerous whale-watching tours are run at that time. For the rest of the year, your only (slim) chance to see marine mammals is to go on an expensive fishing trip. Although almost 30 species of whales and dolphins have been recorded off Bermuda, the islands are located at the edge of the Sargasso Sea, an area of the Atlantic Ocean with very low productivity, so most of these species appear to be rare or accidental visitors. Encountered a bit more often are Short-finned Pilot Whale, Sperm Whale (in spring), and Pantropical Spotted, Atlantic Spotted, Spinner, Striped, Clymene, Rough-toothed, and offshore Common Bottlenose Dolphins. You might get a better chance of seeing all these species if you get onboard a cruise or cargo ship going from the U.S. to Bermuda, as your route would cross the more productive waters in and around the Gulf Stream.

PART III.
SPECIES-
FINDING
GUIDE

TECHNICAL NOTES

This book includes mammals of all 50 U.S. states, Canada, and Greenland. Although biogeographically Hawaii is a part of Polynesia, the only two native Hawaiian mammals that are not completely marine are of American origin, so Hawaii is included, but various U.S. territories in the Pacific are not. Neither are Puerto Rico, the Bahamas, and U.S. territories in the Caribbean, because their mammalian fauna is West Indian rather than North American. The Bermuda Islands are included because their native mammalian fauna is of North American origin. In a few cases I included good locations outside that area but close to its borders.

I do not follow the recent trend of replacing the term *subspecies* with *PSC species*, where PSC stands for *phylogenetic species concept*. Attempts to introduce PSC into scientific systematics have led to a chain reaction of indefinite splitting of taxa for the sole purpose of improving the authors' credentials. The list below mostly follows the "Revised checklist of North American mammals north of Mexico" by Baker et al. (2003), which, to my understanding, is based on reasonably catholic use of BSC (Biological Species Concept). I don't use CSC (Conservation Species Concept) either, as practice has shown that "upgrading" subspecies to full species for the sole purpose of drawing attention to their conservation tends to undermine conservation rather than promote it. I updated the taxonomy as well as I could. Since there are no standard English names, in cases where there was a choice of equally popular alternatives I chose the ones I liked the most. It was impossible to mention all subspecies, but I included some information on the more distinctive and/or interesting ones.

I mention only those introduced species that are widespread and well-established, or of taxonomic interest. There are lots of African and Indian game species on the loose in Texas, monkey bands in

Florida, and other exotics scattered at numerous locations around the continent. I don't think that, for example, seeing introduced Tahr at Hearst Ranch is a valid substitute for seeing it in the Himalayas. To me, chasing recently released exotics is not very different from a trip to a safari park.

Note that this part of the book is divided into chapters based on convenience, not science. Zoologically, pinnipeds should be included in carnivores, jumping mice are not closely related to other mice, and so on.

If I say that some species can be reliably seen at a certain location, I mean that your chances of seeing it there in one full day (or night, whatever is appropriate) of active searching are above 50 percent.

Whenever possible, I provide tips for finding each species yourself, rather than taking a guided tour. Also whenever possible, I provide tips for observing rather than trapping. Unless stated otherwise, "trapping" refers to using Sherman or Elliot traps (metal live traps popular among scientists studying small mammals).

The taxonomy of North American mammals constantly changes. For example, Bighorn Sheep of southwestern California used to be considered a separate subspecies called Peninsular Bighorn, but new genetic data caused them to be lumped with Desert Bighorn, a subspecies inhabiting other arid mountains of the Southwest. In cases where different scientists have conflicting opinions on classification, I have tried to mention both versions.

SHREWS AND MOLES

Seeing these secretive animals in the wild takes either dumb luck or a lot of effort. Almost every place in North America has a few species, but even zoologists studying small mammals often have trouble finding them. Shrews and American Shrew Mole are usually caught by pitfall trapping; you have to check traps every two or three hours to extract the animals alive. Sometimes they get caught by Sherman traps, but normally at rates of less than one shrew per thousand trap-nights. You can improve trapping success by keeping Sherman traps in good condition, with very sensitive triggers, and by adding chopped mealworms or earthworms to the bait. Shrews also get caught in accidental "pitfall traps," so if you find, for example, a line of holes dug across the prairie in preparation for installing utility poles, it's a good idea to take a long stick and check if any trapped shrews are hiding under the debris on the bottom of these holes.

Shrews are sometimes seen on night drives, and can occasionally be caught by hand if you jump out of the car fast enough. If you are interested in salamanders or other animals living under logs and rocks, and flip a lot of those during field trips, you might sometimes find shrews as well. Forest shrews and American Shrew Mole can be found by the rustling noise they make in dry leaves, particularly in hardwood forests in late fall or early spring if the weather is very dry. A good way to see tundra shrews in summer is to follow a Domestic Cat on its hunting forays. Cats frequently catch shrews, but don't eat them unless they are starving. Note than many shrew species claimed in literature to be diurnal are in reality predominantly nocturnal; it's just that people see them more often during the day. After a snowfall, look for tunnel-like shrew trails in the fresh snow, particularly where it has fallen on trails and other hard surfaces. Shrews usually have small home ranges, so watching a place with lots of shrew tracks is a good way to see them sooner or later. Short-tailed shrews are good diggers and build networks of tunnels in soft soil.

Moles are even more difficult to observe because most species seldom show up aboveground; virtually all photos of moles looking out of their molehills are of dead animals. Sherman traps don't work very well on moles; you can mail-order live mole traps from Britain, but I haven't tried them. Catching moles with a shovel involves high risk of injuring the animal. If you happen to witness a flood, especially an unusually extensive one, watch the advancing water edge: sometimes you can see moles, shrews, gophers, voles, and other hard-to-see creatures trying to escape. Moles living in rhododendron or azalea thickets forage on the surface more often, probably because they have trouble building tunnels through the dense root tangles.

Short-tailed shrews are venomous, and their bites can be extremely painful, even if they don't break the skin.

NORTHERN SHORT-TAILED SHREW (*Blarina brevicauda*). A widespread and common shrew of the ne. U.S. and se. Canada, where most shrews seen crossing roads or trails belong to this species. Prefers moist habitats near water and hardwood forests with a lot of dry leaves, but can be found almost anywhere. Rather conspicuous and tame, this and the next species are usually the first shrews that novice mammal watchers learn to locate by sound. Often feeds underneath bird feeders at night, particularly if the feeders contain sunflower seed or cracked corn. Common in almost all protected natural areas within its range, including, for example, Algonquin PP (ON), Midewin National Tallgrass Prairie (IL), Great Dismal Swamp NWR (VA/NC), House Mt. SNA (TN), Shenandoah NP (VA), Spring Mill SP (IN), Trap Pond SP (DE), Jamaica Bay WR (NY), Lonsdale Marsh in Pawtucket (RI), Acadia NP (ME), and Prince Edward Island NP (PE).

SOUTHERN SHORT-TAILED SHREW (*Blarina carolinensis*). Widespread and common in forests, meadows, and fallow fields of the se. U.S. This is by far the most frequently encountered shrew in most parts of the interior Southeast; for example, in Big Thicket NPR (TX), Okefenokee NWR (GA), Kisatchie NF (LA), and Francis Marion NF (SC). On rainy summer nights when winged termites and ants emerge from colonies, look for these shrews gorging themselves on easy prey at colony entrances. Two southernmost races were recently (2004, 2006) proposed to be distinct species. **EVERGLADES SHORT-TAILED SHREW** (*B. peninsulae*) is a small form that is widely distributed in cen. and s. FL; look for it at night along Mahogany Hammock Trail in Everglades NP, and along Hickory Trail in Highlands Hammock SP. **SHERMAN'S SHORT-TAILED SHREW** (*B. shermani*) is a very distinctive, dark form confined to a small coastal area in sw. FL. It is probably less conspicuous than other short-tailed shrews; I've spent a lot of time within its range but saw it only once, in Collier-Seminole SP during a hurricane-caused flood.

ELLIOT'S SHORT-TAILED SHREW (*Blarina hylophaga*). Common in almost all habitats in the s. Midwest and OK, particularly in floodplain forests. Forested parts of Tallgrass Prairie NPR (KS) are a good place to see it on early morning walks, especially in winter. Occurs in Sequoyah NWR (OK), Wichita Mts. WR (OK), Hot Springs NP (AR), and Mark Twain NF (MO).

LEAST SHREW (*Cryptotis parva*). This tiny ball of fury is a common but cryptic dweller of grasslands, meadows, and overgrown fields in the e. U.S.; it also occurs in forests and marshes. Often uses vole and cotton rat runways. I have found this shrew under plywood near the headquarters of Fakahatchee Strand PR SP (FL), on levees separating crawfish ponds near Bunkie (LA), under logs in grassy forest clearings in Bond Swamp NWR (GA), and along pond margins in Patuxent Research Refuge (MD). Reportedly common in Panther SF (WV), Land Between The Lakes NRA (TN/KY), Talladega NF (AL), Yazoo NWR (MS), Aransas NWR (TX), Lower Rio Grande Valley NWR (TX), Wichita Mts. WR (OK), Neosho WA (KS), Midewin National Tallgrass Prairie (IL), and Wayne NF (OH).

SMOKY SHREW (*Sorex fumeus*). A common species of the ne. U.S. and se. Canada. Lives in moist forests and shady ravines with a thick layer of leaf litter, and can often be found by sound. Common in most years in the Appalachians at 3,000–4,000 ft. (900–1,200 m) and on the Cumberland Plateau. Listen for it in Great Smoky Mts. NP (TN/NC), at Mt. Pisgah (NC), in Cumberland Gap NHP (TN/KY/VA), and in Colditz Cove SNA (TN). Occurs in Adirondack and Catskill Mts. (NY), in Natchaug SF (CT), at Mt. Greylock (MA), in White Rocks NRA (VT), in Umbagog NWR (NH/ME), and in Grafton Notch SP (ME). Try also moist coniferous forests with lots of rotten logs in Fundy NP (NB), hemlock groves in Algonquin PP (ON), and forest streams in Kejimkujik NP (NS).

PYGMY SHREW (*Sorex hoyi*). Widespread in Canada, interior AK, and the colder parts of the Lower 48 states. The smallest mammal in the East; Appalachian individuals are particularly tiny. This incredibly hardy creature is generally quiet and very difficult to see, although on windless nights you can sometimes hear its excited high-pitched squeaks as it attacks a large insect prey, such as a cicada larva or a caterpillar. I've observed it along streams in the Cades Cove area of Great Smoky Mts. NP (TN), on overgrown lakeshores of Kenai Peninsula (AK), and in a forest clearing in Riding Mt. NP (MB). In November 2013 this shrew was particularly numerous (and active in the middle of a blizzard despite brutal cold) in roadside ditches in Ashuapmushuan WR (QC). Reportedly common in Umbagog NWR (NH/ME), in Medicine Lake NWR (MT), in Allegany SP (NY), in Denali SP (AK), and on the islands of Lake Champlain (VT).

DWARF SHREW (*Sorex nanus*). The smallest mammal of the Rockies and Black Hills, it inhabits coniferous forests and rocky tundra at high elevations. It can sometimes be found by flipping rotten logs or by watching vole and pika runways between boulders. I've seen it on moist slopes in Valles Caldera NPR (NM), along the timberline on Mt. Evans (CO), and near hollow stumps in recently burned areas in Custer SP (SD). A similar species, **CARMEN MOUNTAIN SHREW** (*S. milleri*), occurs just a few miles south of the U.S./Mexico border, and might be present at higher elevations of Big Bend NP (TX). Try looking for it in pine-oak forests with lots of dry leaves and rotten logs.

INYO SHREW (*Sorex tenellus*). A little-known endemic of the Sierra Nevada; recently found also in Great Basin NP (NV). Occurs on dry slopes, in gulches, and in fir forests above 5,000 ft. (1,500 m). It is apparently easier to find by spotlighting on warm midsummer nights than it is to trap. I've seen it hiding under a log in Ancient Bristlecone Pine Forest in Inyo NF, crossing a road northwest of Mono Lake, and eating a moth on a road shoulder near Tioga Pass (all in CA).

Dwarf Shrew weighs just ¹/₁₆ oz. (2 to 4 g), but it remains active on the high peaks of the Rockies throughout the winter.

ORNATE SHREW (*Sorex ornatus*). Widespread in s. and cen. CA, particularly in wetlands and riparian corridors, but uncommon and difficult to see. Occurs in waterfall spray zones in Yosemite Valley, in oak groves in Mt. Diablo SP, in riverside thickets in Caswell Memorial SP, in marshy grasslands in Vic Fazio Yolo WLA, in shady ravines in Big Sur area, and around vernal pools in Carrizo Plain NMT. The small, dark-colored subspecies that used to occur around Buena Vista Lake is listed as endangered in the U.S.; the lake has been drained and this subspecies might be extinct, but you can try looking for it in Kern NWR and in Tule Elk SR. Three other dark-colored subspecies, also rare and threatened by habitat loss, live in salt marshes around San Francisco Bay and around Los Angeles; look for them in Don Edwards San Francisco Bay NWR, in Palo Alto Baylands Nature PR, Upper Newport Bay ER, and Bolsa Chica ER during winter high tides, when the shrews sometimes climb low shrubs to escape the flooding. The northernmost of these three races is particularly distinctive and has been considered a separate species, **SUISUN SHREW** (*S. sinuosus*).

The best place to look for it is the loop part of Lower Tubbs Island Trail in San Pablo Bay NWR, where some individuals have gorgeous silver-black fur. It also occurs in Grizzly Island WLA. Yet another dark-colored subspecies, this one of remarkably large size, occurs on Catalina I. It is known from just a few specimens and sightings; seems to prefer moist canyons.

ROCK SHREW (*Sorex dispar*). Also known as **LONG-TAILED SHREW**, this little-known animal occurs around rocky outcrops, on shaded talus slopes, and along boulder-filled streambeds in the Appalachians. It is adapted to moving through narrow spaces, and looks a bit lizard-like in its movements. Although largely subterranean and difficult to find, it can sometimes be glimpsed on winter nights when it runs across snow patches between rocks. Try to find its tracks in the snow during the day, then pick a well-used trail and watch it at night. I've seen it near Dark Hollow Falls in Shenandoah NP (VA), near the upper end of Chimney Tops Trail in Great Smoky Mts. NP (TN), and on a rocky slope in Port-Daniel WS (NB). Reportedly common at Craggy Dome (NC) and in New River Gorge (WV). A smaller shrew living in the Maritimes has been described as a separate species, **GASPÉ SHREW** (*S. gaspensis*). Look for it along Les Lacs Trail in Forillon NP (QC), and along small forest streams in Cape Breton Highlands NP (NS).

AMERICAN WATER SHREW (*Sorex palustris*). This large, semiaquatic shrew is widespread in colder parts of N. America, and in mts. south to AZ and AL along streams, in marshes and bogs, and around overgrown lakes. Recently (2014) found to be three different species: **CORDILLERAN WATER SHREW** (*S. navigator*) in Western mts. and on Vancouver I. (BC); **BOREAL WATER SHREW** (*S. palustris*) in lowlands of interior Canada and n. Midwest; and **EASTERN WATER SHREW** (*S. albibarbis*) in e. Canada and the e. U.S. Look for it along Trans-Labrador Hwy. (NL); along meadow streams in Yellowstone NP (WY), North Cascades NP (WA), Yosemite NP (CA), Umbagog NWR (NH/ME), Purgatory Chasm State Reservation (MA), Red Rock Lakes NWR (MT), and City of Rocks NR (ID); in tamarack bogs near Churchill (MB) and in Whiteshell PP (MB). A rare form, sometimes considered a full species, **GLACIER BAY WATER SHREW** (*S. alaskanus*) lives in mossy forests around Gustavus (AK); look for it in summer by walking very quietly along small streams.

MARSH SHREW (*Sorex bendirii*). A rare nocturnal species of the Pacific Northwest, where it occurs in forest wetlands (particularly in skunk cabbage bogs) and along streams. Largely aquatic, it is very difficult to find. In many nights of spotlighting in good habitat I saw it only once, in Devils Elbow SP (OR), as it hunted for stonefly larvae in a shallow forest creek. Reportedly common in Conboy Lake NWR (WA). A white-bellied subspecies occurs on Olympic Peninsula (WA).

BAIRD'S SHREW (*Sorex bairdi*). A small shrew known only from nw. OR, where it is rare and seldom seen. Prefers dense, moist Douglas fir forests with lots of rotting logs, but sometimes lives in lush coastal meadows. A shrew of this species used to live at the northern entrance to Sea Lion Caves, but I didn't see it during my most recent visit. Occurs in Devils Elbow SP, where you can sometimes see it in dark ravines as it runs along well-decomposed logs on foggy winter mornings. Possibly conspecific with Montane Shrew.

PACIFIC SHREW (*Sorex pacificus*). A shy, little-known nocturnal species of the Pacific Northwest. Occurs along overgrown forest streams, in moist forests with lots of rotten logs, and in alder groves. Relatively common in Devils Elbow SP (OR), where you can usually see it at least once in two or three nights of spotlighting. Reportedly occurs around Ross Lake in North Cascades NP (WA). The inland subspecies is very rare; look for it in Trinity Alps (CA) around shaded seepages in dense forest. Listed as threatened in Canada.

FOG SHREW (*Sorex sonomae*). A rare species of the Pacific Northwest, this cute shrew inhabits moist forests, lush meadows, piles of dry seaweed on rocky beaches, and sometimes dense shrubs along small streams on chaparral-covered slopes. I have seen it at dawn on a trail in Muir Woods NMT (CA), and under a rotten log in Devils Elbow SP (OR). Reportedly common in King Range NCA (CA).

MONTANE SHREW (*Sorex monticolus*). Widespread and common in moist woodlands, floodplain meadows, riparian corridors, and wet mountain tundras in AK, w. Canada, and at higher elevations in the w. U.S. Search for it in spring, when grass cover is minimal and areas of shrew activity can be found by looking for tracks on remnant snow patches. Good places include alpine meadows of the Sierra Nevada, wet areas on the floor of Valles Caldera NPR (NM), tundra streams in Denali NP and Denali SP (AK), and almost any high-elevation wetland in CO. In some years this shrew is abundant in September on the summit plateau of Grand Mesa (CO). A distinctive race from s.-cen. NM, described as a new species, **NEW MEXICAN SHREW** (*S. neomexicanus*) can be found in forest gulches above 9,000 ft. (2,700 m) in Capitan Mts. (NM) and at Sandia Crest (NM).

VAGRANT SHREW (*Sorex vagrans*). Common in marshes and wet meadows from s. BC to n. CA and UT, particularly in the vicinity of vole colonies and at higher elevations. Relatively easy to find on Vancouver I. (BC), in Burns Bog near Vancouver (BC), in Crater Lake NP (OR), in City of Rocks NR (ID), in National Bison Range (MT), in Lost Creek SP (MT), in Turnbull NWR (WA), in Fish Springs NWR (UT), and along meadow streams at high elevations in the Sierra Nevada, for example, at Tuolumne Meadows in Yosemite NP (CA). Two rare, distinctive races inhabit salt marshes around the southern part of San Francisco Bay and along the eastern shore of Monterey Bay; look for them during winter high tides in Don Edwards San Francisco Bay NWR, Palo Alto Baylands Nature PR, and Elkhorn Slough NR (all in CA).

MASKED SHREW (*Sorex cinereus*). A common and widespread shrew of the colder parts of N. America, particularly abundant in AK and much of Canada. It used to be called **COMMON SHREW**, but this name has been abandoned to avoid confusion with Eurasian species *S. araneus*. In the mountains it occurs south to NM and AL. An active, valiant, relentless predator, it is the only northern shrew regularly seen during night drives. It generally prefers moist areas, but can be found anywhere from city parks, such as the National Arboretum in Washington, DC, and the grounds of Bronx Zoo (NY), to glacial moraines in BC, tundra streams in Gates of the Arctic NP (AK), meadows around Gustavus (AK), prairies in Lostwood NWR (ND), forest bogs in Voyageurs NP (MN), woodland patches in s. WI, marshy meadows in Yellowstone NP (WY/MT/ID), stream banks in Arapaho NWR (CO), and grassy lakeshores in Umbagog NWR (NH/ME). Particularly common on Newfoundland; look for it in Gros Morne NP (NL). Two distinctive races have been described as full species: **MARYLAND SHREW** (*S. fontinalis*) from the Chesapeake Bay area, which is common, for example, in Trap Pond SP (DE) and Blackwater NWR (MD), and might occur as far north as Belleplain SF (NJ); and strikingly bicolored **ST. LAWRENCE ISLAND SHREW** (*S. jacksoni*), which is reportedly common along coastal tundra streams on St. Lawrence I. (AK).

BARREN GROUND SHREW (*Sorex ugyunak*). Occurs in lowland tundra from AK to mainland NU. Common around Barrow (AK) and Inuvik (NT); look for it under driftwood or wooden construction debris on densely vegetated hillsides near wetlands. Sometimes considered a subspecies of Masked Shrew. The slightly larger race inhabiting marshes and tundra-covered dunes on Pribilof Is. (AK) is often considered a separate species, **PRIBILOF ISLANDS SHREW** (*S. pribilofensis*).

PRAIRIE SHREW (*Sorex haydeni*). Also known as **HAYDEN'S SHREW**, it is an uncommon resident of wet prairies, meadows, and sometimes open woodlands in the northern part of the shortgrass prairie zone. I've seen it during night drives on the lands of the Blackfeet Tribe (MT). Known to occur in Wind Cave NP (SD), Huron Wetlands (SD), Theodore Roosevelt NP (ND), and Grasslands NP (SK).

PREBLE'S SHREW (*Sorex preblei*). A rare, little-known shrew of arid plains and high-elevation forests and marshes in the nw. U.S. and in NV; there are also a few records in the mountains of the Southwest. I have found it only once, while spotlighting on horseback in shortgrass prairies near Duck Lake on the lands of the Blackfeet Tribe (MT). A few shrews I've seen crossing the road during night drives on Hwy. 95 south of Jackpot (NV) probably belonged to this species, but I've never been able to catch one and check. Reportedly occurs in Lassen Volcanic NP (CA), Hells Canyon NRA (OR/ID), and Sheldon NWR (NV).

MOUNT LYELL SHREW (*Sorex lyelli*). One of the rarest land mammals of N. America, this agile hunter is known only from meadow wetlands and sagebrush flats of the cen. Sierra Nevada above 7,000 ft. (2,100 m). It can occasionally be seen on summer night drives as you approach Sonora Pass or Tioga Pass (CA), but I've found it only twice in many nights of driving.

OLYMPIC SHREW (*Sorex rohweri*). A recently (2007) described species, known from lowland wetlands of the Olympic Peninsula (WA), and from Burns Bog on the southern outskirts of Vancouver (BC). I think the shrew I once saw near Lake Ozette Campground in Olympic NP (WA) belonged to this species, but it's difficult to be sure.

SOUTHEASTERN SHREW (*Sorex longirostris*). Widespread but difficult to find in wetlands, open forests, and overgrown fields of the se. U.S. I have found it under a rock slab in mixed forest at Whiteside Mt. (NC), under a rocky overhang in Colditz Cove SNA (TN), and in an old harvest mouse nest in Bond Swamp NWR (GA). Reportedly common in Homochitto NF (MS) and in Congaree NP (SC), but I've never seen it there. A rare, distinctive race occurs in Great Dismal Swamp NWR (VA/NC), where it sometimes can be found in roadside ditches (wait for a snowfall and look for places with shrew tracks).

EURASIAN LEAST SHREW (*S. minutissimus*). Formerly known in N. America as **YUKON SHREW** or **ALASKA TINY SHREW** (*Sorex yukonicus*), this minuscule creature occurs in streamside habitats of interior AK, where it somehow manages to survive and remain active in winter when air temperatures drop below −70°F (−57°C). It is believed to be very rare, but might in fact be common and widespread, just difficult to trap. I have seen it once in a willow-lined creek bed north of Polychrome Glaciers in Denali NP (AK).

ARCTIC SHREW (*Sorex arcticus*). A widespread but difficult-to-find species of boreal wetlands. Most common in taiga bogs and spruce-tamarack swamps of inland Canada; rare and local in the n. Midwest. Look for tiny piles of beetle parts near old muskrat lodges; they are a sign that a shrew might be living inside. On summer nights it is sometimes relatively easy to see in Wood Buffalo NP (NT/AB); occurs also in Voyageurs NP (MN), Medicine Lake NWR (MT), and Pukaskwa NP (ON).

MARITIME SHREW (*Sorex maritimensis*). A rare, somewhat mysterious endemic of bogs and marshes of the Canadian Maritimes; possibly a subspecies of Arctic Shrew. I have caught it once in a coastal marsh in Fundy NP (NB) after it was snatched and then dropped half-dead by a red fox I was watching. Resuscitating it was an interesting experience.

TUNDRA SHREW (*Sorex tundrensis*). Widespread in densely vegetated upland tundra in AK, n. YT, and far n. NT. Common along Teller Hwy. (AK), in Barrow area (AK), and around Inuvik (NT), and can sometimes be found by listening for high-pitched sounds coming from lemming burrows (don't stick your hand in there without a glove: lemmings bite!). Possibly a subspecies of Arctic Shrew.

ARIZONA SHREW (*Sorex arizonae*). Occurs in dry pine-oak forests of extreme s. AZ and sw. NM, where it can be located by the noise it makes in leaf litter. The only shrew of this species I've ever seen was in a dry rocky streambed in Madera Canyon (AZ); when I spotted it late at night, it was busy dispatching a large centipede.

MERRIAM'S SHREW (*Sorex merriami*). Uncommon in dry grasslands, sagebrush deserts, and dry pine forests of the w. U.S. In some places, such as in Great Sand Dunes NP (CO), it inhabits sand dunes and can be found early in the morning by following its tracks to its shelter, often in a kangaroo rat burrow or under a shrub. On moonless nights, look for it in Sagebrush Vole colonies in Thunder Basin NG (WY), in sandy flats along the southern shore of Mono Lake (CA), in grasslands in and around City of Rocks NR (ID), and in dry juniper woodlands in Hells Canyon NRA (OR/ID). On arid plains these shrews sometimes follow swarms of Mormon crickets in the same way Gray Wolves follow migrating Caribou herds. If you encounter a swarm, look for shrews scurrying along its tail edge.

TROWBRIDGE'S SHREW (*Sorex trowbridgii*). The most frequently seen shrew in old-growth forests (except redwood-dominated stands) of w. coastal U.S., east to the Sierra Nevada; often climbs trees and can be found in tree hollows. Occasionally seen at dusk on forest trails in Santa Cruz Mts., particularly in Forest of Nisene Marks SP, and during night drives along the road to Mineral King in Sequoia NP (both in CA). Common in Burns Bog near Vancouver (BC), in Olympic NP (WA), in Devils Elbow SP (OR), and in Redwood NP (CA).

CRAWFORD'S DESERT SHREW (*Notiosorex crawfordi*). The commonest shrew of the desert Southwest, it occurs in all kinds of arid habitats around dry logs, woodrat nests, cholla groves, and piles of dead leaves. Occasionally seen during night drives, for example, in Joshua Tree NP (CA), in Organ Pipe Cactus NMT (AZ), and at Granite Gap (NM). Reportedly common in Lower Rio Grande Valley NWR (TX).

COCKRUM'S DESERT SHREW (*Notiosorex cockrumi*). A recently (2003) described species from mesquite deserts of se. AZ; reportedly occurs in Leslie Canyon NWR. I have never seen it in the U.S. despite much searching, but found one in Mexico near Nacori Chico (Sonora) in a pile of dry agave leaves. One or two smaller species of desert shrews are known from very recent fossils and might still occur somewhere in the Southwest.

AMERICAN SHREW MOLE (*Neurotrichus gibbsii*). Common in second-growth redwood forests and alder groves of the Pacific Northwest, south to Monterey. It is shrewlike in habits, but noisier, and can easily be caught by hand if located by sound. I often saw these small moles in the woods around my home when I lived in Santa Cruz Mts. (CA). Skagit WLA (WA) and Alfred A. Loeb SP (OR) are also good places to listen for them.

STAR-NOSED MOLE (*Condylura cristata*). This unique creature is widespread in the ne. U.S. and e. Canada, and present at higher elevations in the Appalachians and along the Atlantic Coast south to GA. Inhabits moist woods and meadows, where it usually lives near small streams, ponds, and swamps. It can feed on the surface, underground, and underwater. Although generally elusive, it is a bit more likely to be seen on cold nights in late fall or early spring, when the soil is frozen but there is little or no snow on the ground. Look for it in small forest ponds in Ottawa NWR (OH), along swamp boardwalks in Great Swamp NWR (NJ), around the first bridge on Mekinac Trail in La Mauricie NP (QC), and in Esther Currier WMA (NH). Reportedly common in Sunkhaze Meadows NWR (ME) and in Great Meadows NWR (MA).

BROAD-FOOTED MOLE (*Scapanus latimanus*). Common in much of CA and locally in s.-cen. OR in areas with soft soil and sparse grass. Abundant in Santa Cruz Mts. (CA), but very rarely leaves its burrows. This is, however, the only mole I've ever seen cross a trail (in Ishi Wilderness, CA). Try also Kruse Rhododendron SR (CA), preferably on a foggy night. An isolated population lives on the nw. side of Mono Lake (CA).

COAST MOLE (*Scapanus orarius*). Common in open forests and dry meadows of the Pacific Northwest. Sometimes forages aboveground in places with dense shrubs and lots of leaf litter, mostly at night when there is heavy fog and light drizzle. Azalea SR near Arcata (CA) is a good place to look for it. Common in coastal meadows in King Range NCA (CA).

TOWNSEND'S MOLE (*Scapanus townsendii*). This large, beautiful mole is common in wet meadows and moist forests in the Pacific Northwest. During winter storms it can sometimes be seen on the surface. Easy to find but difficult to see in and around Redwood NP (CA), in shady gulches in King Range NCA (CA), and in Ridgefield NWR (WA). If you happen to see a large tree fall, check the area under the roots immediately—sometimes you can find moles there.

HAIRY-TAILED MOLE (*Parascalops breweri*). Widespread in woodlands and overgrown fields in s. ON, the ne. U.S., and at higher elevations in the Appalachians. This mole regularly forages aboveground at night and sometimes on overcast days. Look for it in rhododendron groves on the summit of Roan Mt. (TN), in azalea thickets on the summit of Gregory Bald in Great Smoky Mts. NP (TN/NC), and in Rhododendron SP (NH). Common at lower elevations in Adirondack Mts. (NY), on the shores of Lake Champlain (VT/NY/QC), and in Cuyahoga Valley NP (OH).

EASTERN MOLE (*Scalopus aquaticus*). Locally common in areas with soft soil throughout the e. U.S.; usually absent from coastal lowlands, floodplains, and high mountains. This is the most numerous and widespread mole in N. America, but also the most strictly subterranean. Even if it does show up on the surface, it does so on moist, sometimes rainy nights when the leaf litter doesn't make much noise. In places with rocky substrate under a thin layer of soft soil it can occasionally be found under rocks and large logs. Wichita Mts. WR (OK), Frozen Head SP (TN), Wertheim NWR (NY), Pachaug SF (CT), Chesapeake and Delaware Canal WA (DE), and many vacant lots around Detroit (MI) have good populations. The large, blonde-colored southern form is even more difficult to find. In the many years I've lived in the South, I have seen it on the surface only once: it was hunting for winged termites under the parking lot light at Kirby Trailhead in Big Thicket NPR (TX), on a warm, moist, overcast spring night. It can reportedly be found aboveground in low-lying parts of Ocmulgee NMT (GA) on very rainy nights.

BATS

Almost all North American bat species can be found in caves at least sometimes, but show caves tend to lose their bat colonies because of disturbance, and nontouristy caves (as well as abandoned mines) are increasingly gated either under pretense of protecting public safety or in futile attempts to prevent the spread of white-nose syndrome. Although closing access to caves doesn't prevent the spread of the syndrome, limiting disturbance to bat caves is generally a good thing. It is also a good habit to always thoroughly wash your clothes and shoes after visiting caves, or even to buy new shoes and clothes for caving if you move between states. Each cave is a unique ecosystem, so carrying any microorganisms between caves might lead to serious damage. If a cave is gated, you can often see bats near the entrance, particularly at night. Entering undeveloped caves, abandoned mines, and uninhabited houses can be dangerous—attempt at your own risk.

Information about the locations of undeveloped caves is seldom made public, supposedly to prevent vandalism. The best way to get better access to little-known local caves is to join the caving club at the local university, if there is one. Joining such a club (usually called a *grotto*) would also give you a chance to participate in bat counts and other interesting research.

Other good methods for finding bats include regularly checking hollow trees, artificial nest boxes, and abandoned buildings (you can use a dentist's mirror with a built-in light to look into small openings). During the day, bats roosting inside building walls, under roofs, in dry palm fronds, or in clumps of Spanish moss can be found by the sounds they make, the smell, or the droppings underneath. Many bats have not just daytime roosts, but also nighttime roosts, which are often more exposed (undersides of bridges, highway culverts, rock overhangs, arches, cave entrances). Trapping bats is usually labor-intensive and can be harmful to the animals, especially if you don't have proper training. Note that possessing mist nets without a permit is illegal in the U.S., and that handling bats is unsafe if you haven't been vaccinated against rabies. A great way to see many bat species up close is to join bat-catching tours led by Fiona Reid (www.fionareid.ca) or training workshops run by Bat Conservation International.

Bats can sometimes be identified in flight, but without a bat detector such identifications are seldom reliable, except for very large or bright-colored species, or those seen and photographed at hummingbird feeders. You can try to attract bats by constructing a bat house (see www.batcon.org for detailed manuals).

GHOST-FACED BAT (*Mormoops megalophylla*). Common in the tropics but rare in N. America, where it has been recorded in extreme s. AZ and TX. Roosts in large colonies in caves and mines in arid areas. Reportedly occurs in winter in Frio Cave near Concan, and in summer in Haby Cave and Valdina Farms Sinkhole in Medina Co. and in Webb Cave in Kinney Co.; present in Big Bend NP (all in TX).

CALIFORNIA LEAF-NOSED BAT (*Macrotus californicus*). Roosts in caves, mines, and sometimes abandoned buildings in southern parts of CA, NV, and AZ; forages low above ground in deserts and arid mountains. Occurs in Organ Pipe Cactus NMT (AZ), where it can reportedly be seen drinking from small seeps; in small caves around Salton Sea (CA); and in abandoned mines in Beaver Dam Wash NCA (UT). This species is extremely sensitive to disturbance and can abandon a colony after just one careless visit by people, so if you find it, please view from afar and don't use camera flash. I've never seen it in the U.S., only in nw. Mexico where it is very common in abandoned mines.

JAMAICAN FRUIT-EATING BAT (*Artibeus jamaicensis*). This large, conspicuous bat is abundant in the tropics but very local in N. America, where it occurs only in w. Florida Keys. Roosts in buildings and hollow trees. Look for it at night around fruiting fig trees in Key West (FL), particularly in Key West Botanical Garden.

CUBAN FLOWER BAT (*Phyllonycteris poeyi*), **BUFFY FLOWER BAT** (*Erophylla sezekorni*), and **CUBAN FIG-EATING BAT** (*Phyllops falcatus*) are W. Indian species, known in N. America from one specimen each, caught near Key West (FL). They roost in caves and abandoned buildings and feed in forests and gardens. Look for them around flowering and fruiting trees in Key West Botanical Garden and elsewhere in town.

Jamaican Fruit-eating Bat occurs in North America only in the lower Florida Keys.

Cuban Fig-eating Bat.

MEXICAN LONG-TONGUED BAT (*Choeronycteris mexicana*). Roosts in summer in caves and mines (usually near the entrances) in extreme s. CA, AZ, and NM. This bat is very shy and difficult to see well. I have observed it hovering around saguaro flowers in Saguaro NP (AZ), and found a few in an abandoned mine in Peloncillo Mts. (NM). The easiest way to see it is to watch the hummingbird feeders in Ramsey and Cave Creek Canyons (AZ) at night.

LESSER LONG-NOSED BAT (*Leptonycteris yerbabuenae*). Common in summer in s. AZ and sw. NM. Roosts in mountain caves and mines, such as Colossal Cave (AZ). Feeds on nectar of flowering cacti and agaves; look for it around flowering cacti in the Arizona-Sonora Desert Museum and in Las Cienegas NCA (AZ). Often visits hummingbird feeders on summer nights, for example, in Ramsey Canyon (AZ), and will find new ones within a few hours—just put them up in a forested canyon. Listed as endangered in the U.S. (as *L. curasoae yerbabuenae*).

GREATER LONG-NOSED BAT (*Leptonycteris nivalis*). Sometimes called **MEXICAN LONG-NOSED BAT**, it occurs in summer around flowering cacti and agaves in Big Bend NP (TX) above 5,000 ft. (1,500 m); recorded also in Animas Mts. (NM). Roosts in caves and mines. Can be attracted to hummingbird feeders, especially if you put them out for more than one night. Listed as endangered in the U.S.

HAIRY-LEGGED VAMPIRE BAT (*Diphylla ecaudata*). A tropical species, known in N. America from one specimen collected in a railroad tunnel near Comstock (TX) in 1967. It doesn't normally occur anywhere near the U.S./Mexico border, and might never be found in the U.S. again, unless global warming gets really bad. Feeds on bird blood.

LITTLE BROWN MYOTIS (*Myotis lucifugus*). The most widespread bat in N. America (except the southernmost parts), and the only species in AK north of the se. Panhandle. Roosts in buildings and small caves in summer, and in larger caves and mines in winter. In summer, easy

Little Brown Myotis is one of the species hit particularly hard by white-nose syndrome.

to see in Maquoketa Caves SP (IA), in Wyandotte Caves (IN), and in caves near Ledge View Nature Center (WI). Small roosts are often present in lava tubes of Mt. St. Helens NMT (WA), and in small caves in Mammoth Cave NP and Natural Bridge SP (both in KY); in summer these bats sometimes roost at night at the main entrance to Mammoth Cave itself. Breeds in summer in caves in Rifle Falls SP (CO); although the main colony there is inaccessible, you can often see single bats emerging from smaller caves or roosting in larger ones, especially on weekday nights. Present year-round in Endless Caverns (VA), Gap Cave, Appalachian Caverns, and sometimes Bristol Caverns (all in TN); in winter also in

Morrill Cave (TN). From mid-September to late November it can be reliably seen in Bonnechere Caves (ON). Colonies exist in bat houses in Moore SP (MA) and Canoe Creek SP (PA), as well as in the church near the latter park, and in United Methodist Church in Tranquility (NJ). A few bats are often present in small caves and rocky niches in Purgatory Chasm State Reservation (MA). The dark form inhabiting the Pacific Northwest can sometimes be seen in winter in Oregon

Caves NMT (OR). Very rare near the northern limit of its range; in AK I've seen it only once, in Matanuska-Susitna Valley. Highly susceptible to white-nose syndrome, and may soon disappear from much of the continent. Listing as endangered is expected in Canada.

FRINGED MYOTIS (*Myotis thysanodes*). Widespread but uncommon in the w. U.S. Roosts in caves, mines, and buildings; forages over deserts and in open woodlands. Can be seen in hollow sequoias in Kings Canyon NP (CA), in abandoned mines in Death Valley NP (CA), in rock crevices in Carlsbad Caverns NP (NM), and in abandoned prairie homesteads throughout w. NE and ne. CO. Occasionally roosts in boulder caves in Pinnacles NP (CA). The dark-colored northwestern form occurs in hollow redwoods in Redwood NP (CA) but is a bit uncommon; usually you have to check dozens of redwood hollows before you find it.

LONG-EARED MYOTIS (*Myotis evotis*). Widespread in the West, but little known and difficult to find. Roosts under loose bark, in hollow trees and logs, small caves, rock crevices, rodent burrows, and buildings in forested areas. Occurs in redwood snags in Santa Cruz Mts. (CA), in buildings and cliffs in Fish Springs NWR (UT), in rock crevices in City of Rocks NR (ID), and in vacant bird nest boxes around Boulder (CO). Occasionally roosts in boulder caves in Pinnacles NP (CA). A few are present in summer in caves in Rifle Falls SP (CO); although the main colony there is inaccessible, you can sometimes see single bats emerging from smaller caves or roosting in larger ones, especially on weekday nights. Occasionally lives in bat houses of "rocket box" type (see Bat Conservation International website for instructions on how to build them). The rare shorter-eared form inhabiting coastal forests from Glacier Bay NP (AK) to Olympic Peninsula (WA) has previously been known as **KEEN'S MYOTIS** (*M. keenii*); it can be found in Douglas fir hollows in Cathedral Grove on Vancouver I. (BC).

ARIZONA MYOTIS (*Myotis occultus*). This little-known bat is widespread but uncommon in lowland deserts and open woodlands of extreme se. CA, AZ, and NM. Roosts in caves and mines. Look for it on summer nights under highway bridges in Marble Canyon area (AZ) and around Blythe (CA), in winter in lava tubes of El Malpais NCA (NM), and year-round in abandoned mines in the eastern part of Sonoran Desert NMT (AZ).

LONG-LEGGED MYOTIS (*Myotis volans*). A common forest bat in the West north to AK Panhandle, particularly in canyons and mountain forests. Roosts in caves, rock crevices, hollow ponderosa pines, and abandoned buildings; winters in caves and mines. A few can often be seen on winter tours of Oregon Caves NMT (OR). Roosts in summer in small caves and rock niches in Capitol Reef NP (UT), in spring and fall in hollow sequoia trees in Yosemite NP (CA), and from spring to fall in cliffs and abandoned buildings around Staley Reservoir near Fish Springs NWR (UT).

INDIANA MYOTIS (*Myotis sodalis*). Often called simply **INDIANA BAT**, it is widespread in bottomland forests of e. U.S. Found mostly in hollow trees or under loose tree bark in summer and in cold caves in winter. Although all caves with large colonies are closed in winter, you can sometimes see a few bats in late fall or winter during tours of Mammoth Cave (KY) and Wyandotte Caves (IN). In spring, look for tight clusters of these bats in Gap Cave in Cumberland Gap NHP (TN/KY/VA). Used to be easy to see in small caves of Spring Mill SP (IN), but these caves have been closed in response to white-nose syndrome outbreak (you can still see this bat in hollow trees there in summer and early fall). A colony (mixed with more numerous Little Brown Myotis) exists in a church adjacent to Canoe Creek SP (PA), where you can ask park rangers for directions to the church, or watch the bats emerge from a smaller colony in the bat house the park has built as an alternative shelter. Summer roosts in hollow trees can be found in Cypress Creek NWR (IL); inquire about the possibility of accompanying local scientists monitoring this species. Indiana Myotis is highly susceptible to white-nose syndrome, and may soon disappear from much of its range. Listed as endangered in the U.S.

SOUTHWESTERN MYOTIS (*Myotis auriculus*). Common in pine-oak forests, upland deserts, and rocky canyons of s. AZ and sw. NM. Roosts in hollow trees, in rock crevices, and at cave entrances. I have found single bats of this species in ancient ruins in Gila Cliff Dwellings NMT (NM) and inside an old cow skull in San Bernardino NWR (AZ). Reportedly common in San Pedro Riparian NCA (AZ).

NORTHERN MYOTIS (*Myotis septentrionalis*). Sometimes called **NORTHERN LONG-EARED BAT**, it is widespread in upland forests of Canada east of the Rockies and the e. U.S.; rare in the Southeast. Roosts mostly in hollow trees and buildings in summer and in crevices in cave ceilings in winter, but I have seen a few in late spring in Onondaga Cave (MO) and in early fall in Bluespring Caverns (IN). Small numbers are usually present in winter in Morrill Cave and Gap Cave (both in TN). Occurs in summer in hollow trees at Cape Cod (MA), in Natchaug SF (CT), and on Backbone Mt. (MD). Forages in forests, flying just above the undergrowth. Highly susceptible to white-nose syndrome, and may soon disappear from much of its range; listing as endangered is expected in both U.S. and Canada.

CALIFORNIA MYOTIS (*Myotis californicus*). One of the most common bats in sw. U.S. and along the Pacific Coast north to Haynes (AK). There is a lot of color variation in this species, with the darkest bats occurring in the Pacific Northwest, and the palest ones in the deserts of the Southwest. Roosts in hollow trees in Caswell Memorial SP (CA), in hollow sequoia logs in Kings Canyon NP (CA), in abandoned mines in Death Valley NP (CA/NV), and under rock overhangs in slot canyons of Grand Staircase-Escalante NMT (UT). Often found in isolated buildings in open landscapes, such as in visitor centers and lodges in Petrified Forest NP (AZ), Grand Canyon NP (AZ), and Big Bend NP (TX). Winters in caves and mines, but can often be seen flying in the open on warm winter evenings. Forages over deserts, woodlands, and canyons.

EASTERN SMALL-FOOTED MYOTIS (*Myotis leibii*). An uncommon species of ON, ne. U.S., Appalachian forests, and the Ozark Plateau. Winters in caves, but is difficult to find because it hides in small crevices and under rocks. Used to be easy to see in small caves of Spring Mill SP (IN), but these caves have been closed in response to white-nose syndrome outbreak. Occasionally present in winter in caves of e. TN, usually near entrances. In summer it roosts in rock crevices or under tree bark, for example, in Ouachita Mts. (OK/AR), and in small mica mines in ON. On summer nights, roosting bats can sometimes be seen under highway bridges in Big South Fork NRA (TN/KY), as well as in the "Castle" and in rock crevices in Sleeping Giant SP (CT). Forages mostly in forests.

WESTERN SMALL-FOOTED MYOTIS (*Myotis ciliolabrum*). Widespread but uncommon in arid parts of the West, especially in rocky areas. This is the bat most likely to be seen in ghost towns of CA and NV. Roosts in spring and summer in rock crevices in Mecca Hills WA (CA), in road culverts near Pyramid Lake (NV), in ancient ruins in Mesa Verde NP (CO), in cliffs and buildings in Fish Springs NWR (UT), and under tree bark in burned forests in Yosemite NP (CA). Forages over deserts, prairies, and woodlands; often joins summer swarms of Canyon Bats at the stream crossing near Wolfe Ranch in Arches NP (UT). Winters in caves and mines.

YUMA MYOTIS (*Myotis yumanensis*). One of the most frequently seen bat species throughout much of the West. Roosts in buildings, rock crevices, and under bridges; forages low over water. A large colony lives in a bat house near the headquarters of Clear Lake SP (CA). Another colony (mixed with other species) forms in summer under the pedestrian bridge across Covell Blvd. west of F St. intersection in Davis (CA). Common in abandoned prairie homesteads in se. CO and in small towns along the e. foothills of the Sierra Neveda. Individual bats can sometimes be seen in visitor centers and the inn in Petrified Forest NP (AZ), and often join summer swarms of Canyon Bats at the stream crossing near Wolfe Ranch in Arches NP (UT). The dark northwestern form can often be found at night under road bridges in coastal OR. Winters in caves and mines, but is difficult to find at that time because it squeezes itself into the tiniest cracks in the ceiling.

CAVE MYOTIS (*Myotis velifer*). Locally common in arid lowlands of sw. U.S. Roosts in caves, mines, and under bridges in cracks and old swallow nests, often in the company of Mexican Free-tailed Bat. Pale-colored in TX and OK; dark-colored farther west. Usually present in Gorman Cave (TX), and sometimes in ancient cliff dwellings in Canyon de Chelly and Walnut Canyon NMTs (AZ). Common in abandoned mines around Gallup (NM) and Vidal (CA), but many of these mines are in really poor condition and shouldn't be entered unless you absolutely have to; I'm sure you'll find this species elsewhere if you keep looking. Evening flights can be seen at Frio Cave, Old Tunnel SP, and Eckert James River Bat Cave PR (all in TX), but these caves have many bat species, so you have to be really good at in-flight identification to spot this species among others. On summer evenings you can also watch these bats flying over the Colorado River between Blythe (CA) and Yuma (AZ).

GRAY MYOTIS (*Myotis grisescens*). This highly social bat roosts in caves with running streams from OK and IL to FL and WV; forages over water and in surrounding woodlands. All main colonies are protected, but might soon disappear since this species is susceptible to white-nose syndrome. Mass emergence from large colonies can be seen in summer at Nickajack Cave (TN), Judges Cave near Marianna (FL), and Sauta Cave (AL), but the caves themselves are off-limits. Small groups can often be seen in late fall and winter on regular tours of Mammoth Cave (KY) and in small caves nearby. Look for this bat also at Woods Reservoir Dam at Arnold Air Force Base (TN), and in Wyandotte Caves (IN). Listed as endangered in the U.S.

SOUTHEASTERN MYOTIS (*Myotis austroriparius*). Uncommon in the se. U.S. Roosts in partially flooded caves, buildings, and hollow trees; forages low over water. All major roosting caves have been gated recently, but small groups can be found in caves and rock crevices, for example, in Spring Mill SP (IN) and Hickory Nut Gorge (NC). The commonest bat in caves of FL; mass emergence can be seen in spring and summer at Judges Cave near Marianna (FL)—inquire at Florida Caverns SP. These bats also live in bat houses on Museum Rd. on the University of Florida campus in Gainesville (FL), where they usually emerge earlier than the more numerous Mexican Free-tailed Bats.

SEMINOLE BAT (*Lasiurus seminolus*). Common in summer in se. U.S.; often the most numerous bat species in lowland hardwood forests, live oak groves, and old parks. Roosts in dense clumps of Spanish moss, and can be found by looking for bat droppings underneath trees with lots of Spanish moss "beards." Although brightly colored and exquisitely beautiful, it can be surprisingly difficult to spot even if you know where to look. Forages in forest canopy. Check for it in mahogany trees in the middle of Anhinga Trail parking lot in Everglades NP (FL), along the cypress boardwalk in Highlands Hammock SP (FL), around the parking lot in Audubon SHS (LA), and in the largest trees in Atchafalaya NWR (LA). In late spring it often forages around electric lights near the visitor center at Big Thicket NPR (TX) just after sunset. Occasionally shows up in Bermuda in the fall.

WESTERN RED BAT (*Lasiurus blossevillii*). Widespread in sw. U.S. This species roosts in deciduous trees, particularly along rivers, and is easier to see in flight than roosting. I saw it up close only twice: in the restroom at the Summit Trail parking lot on Mt. Shasta (CA), apparently during fall migration, and in a cottonwood tree in Escalante Petrified Forest SP (UT), where these bats fly around the campground in late spring. Common in forested canyons of se. AZ, where it reportedly roosts in old squirrel nests. Large swarms feed in mid-summer in Gray Lodge Wildlife Area (CA).

EASTERN RED BAT (*Lasiurus borealis*). Common and widespread east of the Rocky Mts.; remains active in winter north to VA. Roosts in trees and sometimes in buildings; hibernates in caves, such as Gorman Cave (TX); in rock crevices, for example, in Hickory Nut Gorge (NC); in bird nest boxes; and on the ground under leaf litter. In the Southeast it prefers to roost in clumps of Spanish moss; look for it in old plantation parks of LA and MS. Forages in forests and clearings. Thanks to its bright coloration, it can be easily recognized in flight if it crosses the road in front of your car or if you manage to catch it in a flashlight beam. Flying bats can often be observed at night at lower elevations in Great Smoky Mts. NP (TN/NC). Some bats migrate south for the winter; look for them in the fall in isolated trees in otherwise treeless coastal marshes of the Gulf Coast, for example, in St. Marks NWR (FL), Bon Secour NWR (AL), Grand Bay NWR (MS), and Lacassine NWR (LA), or in late spring at Sandy Hook in Gateway NRA (NJ). Migrating bats occasionally reach Bermuda in late fall.

HOARY BAT (*Lasiurus cinereus*). This large, beautiful bat has a weird distributional pattern. In summer it is very widespread but uncommon in forested areas; females inhabit the East and migrate north as far as NU, while males occur in the West north to NT. Both sexes winter throughout the s. U.S. in buildings and hollow trees. Summer roosts are in tree branches in dense clumps of leaves or twigs, but sometimes also under loose bark or in bird nest boxes. Forages over water or around streetlights; sometimes catches small bats. With good lighting it can be easily recognized in flight by its large size and unique wing pattern. Occurs in summer near water in Shenandoah NP (VA) and in Caswell Memorial SP (CA), in spring in Hoosier NF (IN), year-round at higher elevations in Big Bend NP (TX), and in the fall in pine forests of Kisatchie NF (LA). Migrating bats sometimes hunt in broad daylight; look for them in September along old canals in e. PA, NJ, and MD. This is the only native land mammal in HI, where the island subspecies can sometimes be seen at dusk flying around Volcano House in Hawaii Volcanoes NP, above the surf in Hilo, and at the bottom of Waimea Canyon. Regularly shows up in Bermuda in the fall, and has crossed the Atlantic all the way to Europe on a few occasions. The HI subspecies is listed as endangered in the U.S.

SOUTHERN YELLOW BAT (*Lasiurus ega*). A common tropical species, very local in N. America. Occurs in Bentsen-Rio Grande Valley SP and Lower Rio Grande Valley NWR (both in TX), where it roosts in dry palm fronds (look for bat droppings underneath palm trees) and forages over water. In Mexico this bat often roosts in large road culverts, so you might try checking them in extreme s. TX as well.

WESTERN YELLOW BAT (*Lasiurus xanthinus*). Uncommon in southern parts of CA, NV, and AZ. Roosts in dry palm fronds and bunches of dry oak leaves; forages over deserts and canyons. Look for its droppings under palm trees in desert canyons, for example, in Mecca Hills WA, Anza-Borrego Desert SP, the southern part of Joshua Tree NP, and Indian Canyons in Santa Rosa and San Jacinto NMT (all in CA).

NORTHERN YELLOW BAT (*Lasiurus intermedius*). Common in coastal se. U.S., but difficult to find. Roosts in dry palm fronds in Bentsen-Rio Grande Valley SP (TX), and in clumps of Spanish moss in Alligator River NWR (NC). Abundant in FL sand scrub, for example, in Ocala NF (FL), where large feeding flocks can sometimes be observed in summer. Another place to see large feeding aggregations on summer evenings is along the Mississippi R. between Natchez (MS) and New Orleans (LA). Also on summer evenings, look for these bats emerging from old squirrel nests in large live oaks on the Louisiana State University campus in Baton Rouge (LA).

SILVER-HAIRED BAT (*Lasionycteris noctivagans*). This beautiful bat is abundant in well-watered forested areas through much of N. America, but rare in the Southeast. Roosts in tree hollows, bird nests, under loose bark (particularly of willow, maple, and ash trees), sometimes also in dense foliage or in rock crevices. Winters in caves and buildings. Often can be found in winter and spring in Onondaga Cave (MO), in summer in rock crevices in Hickory Nut Gorge (NC), and in the fall in old chimneys in Shawnee NF (IL). Single bats sometimes roost on summer nights under river bridges in Natchaug SF (CT) and White Clay Creek SP (PA/DE), in rock niches in City of Rocks NR (ID), and in large tree hollows in old-forest area of Allegany SP (NY).

CANYON BAT (*Parastrellus hesperus*). Formerly known as **WESTERN PIPISTRELLE** (*Pipistrellus hesperus*), this is the smallest bat in N. America. Common in arid parts of w. U.S., usually near water; often forages in broad daylight and returns to roost soon after sunset. Roosts in palm fronds and small rock crevices in desert canyons, for example, in Joshua Tree NP (CA) and in almost every canyon of the Colorado Plateau, from Dinosaur NMT (UT/CO) in the north to Grand Canyon NP (AZ) in the south. It is perfectly adapted to life in rocky habitats; at dawn you can sometimes see these bats disappear in the tiniest rock cracks you'd never even notice otherwise. Also common in coniferous forests of the Sierra Nevada, where in late fall males fly at night along forest roads, calling loudly. The lower reaches of Yosemite NP (CA) are a good place to see these display flights. Large swarms can be seen in summer at the stream crossing near Wolfe Ranch in Arches NP (UT) and in cottonwood groves in Canyonlands NP (UT).

TRICOLORED BAT (*Perimyotis subflavus*). Formerly known as **EASTERN PIPISTRELLE** (*Pipistrellus subflavus*), this bat is the smallest in the East. Widespread and common in forested areas of e. U.S. and adjacent parts of Canada, but rare along the Gulf Coast. Roosts in buildings and hollow trees; winters in caves and buildings; a few can be present in caves in summer. Look for it in Cascade Caverns and Gorman Cave (TX), Forbidden Caverns and Gap and Morrill Caves (all in TN), Wyandotte Caves (IN), Alabaster Caverns SP (OK), Florida Caverns SP (FL), and any large cave in the Ozarks (AR/MO). This tiny

Eastern Pipistrelle can survive all winter covered in dewdrops or frost crystals.

creature is incredibly tough and can survive the entire winter being covered with cold dewdrops or frost, but white-nose syndrome is decimating its populations in the northern part of the range. The dark-colored northern form, which occurs in New England and se. Canada, is becoming very rare because of massive die-offs. Most caves in New England have been closed and gated, but you can still try Howe Caverns (NY), Nickwackett Cave (VT), Morris Cave near Danby (VT), and Warsaw Caves CA (ON) in summer, as well as Bonnechere Caves (ON) in October and November. Listing as endangered expected in Canada.

BIG BROWN BAT (*Eptesicus fuscus*). Common and widespread in much of N. America. Usually roosts in buildings and hollow trees, winters in caves and mines, and forages in all habitats, including cities. A few can be present in caves or fly around the entrances at any time of the year. Look for it in summer and early fall in Mammoth Cave (KY), in caves of Spring Mill SP (IN), in Maquoketa Caves SP (IA), in rock crevices in City of Rocks NR (ID), in caves near Ledge View Nature Center (WI), and in woodpecker holes in Rock Cut SP (IL). In winter, common in Morrill Cave and Appalachian Caverns (both in TN), and occasionally seen during tours of Lewis and Clark Caverns SP (MT). Occasionally roosts in boulder caves in Pinnacles NP (CA). Often found in park offices; ask park rangers, for example, in Cosumnes River PR (CA). The commonest bat in many eastern cities, including Baton Rouge (LA), Knoxville (TN), and Ottawa (ON). In OH it's the most common bat species roosting in old rural churches, particu-

Big Brown Bat is one of the most common bat species, particularly in the East.

larly in Wayne NF. The pale-colored southwestern subspecies can be seen year-round in abandoned uranium mines near Temple Mt. (UT), and in summer in rock crevices in Water Canyon (NM).

EVENING BAT (*Nycticeius hymeralis*). Widespread and common in the Southeast; uncommon in the Midwest. Roosts in buildings and hollow trees; forages over open areas and ponds. Common in abandoned houses in cypress swamps of LA and MS, and in old rural churches from e. TX to SC. In Congaree NP (SC) they often roost under wooden boardwalks on summer nights. Small groups roost in tree hollows in Bastrop SP (TX) and in one of two bat houses in Samuel Houston Jones SP (LA). I have also found these bats roosting on summer nights in road culverts in Mark Twain NF (MO) and under bridges across Buffalo R. between Ponca and Buffalo City (AR).

SPOTTED BAT (*Euderma maculatum*). This alien-looking wonder with an almost mythical status among mammal watchers is widespread in arid parts of the West, but very difficult to find despite having loud flight calls. Don't despair: the countless hours you'll have to devote to finding it will be spent in some of the most beautiful places on Earth. It roosts in deep crevices in canyon walls and cave ceilings, and forages over dry plains, pastures, and forest clearings. I found a small colony in a rock crevice in Mecca Hills WA (CA) in April 2003, but subsequent searches failed to locate that colony. With a good flashlight, this bat can occasionally be seen in flight near tall cliffs in Vermilion Cliffs NMT (AZ), in Paria Canyon-Vermilion Cliffs Wilderness (UT/AZ), in Arches NP and Canyonlands NP (both in UT), and, reportedly, near the bottom of the Grand Canyon (AZ) and in Gunnison Gorge NCA (CO). Some bats roosting in the Grand Canyon fly every night to feed over forest lakes of Kaibab Plateau, 30 to 40 miles to the north. Reportedly roosts in rock crevices in City of Rocks NR (ID).

ALLEN'S BIG-EARED BAT (*Idionycteris phyllotis*). Similar to Spotted Bat but more conventionally colored, this bat is little known and believed to be uncommon. It occurs in sw. U.S., where it roosts in rock piles, mines, and hollow trees; forages over open woodland; and winters in caves and mines. It occurs in abandoned mines in high-elevation portions of Death Valley NP (CA/NV); also present in Cave Creek Canyon, in Agua Fria NMT, and on the North Rim of the Grand Canyon (all in AZ).

RAFINESQUE'S BIG-EARED BAT (*Corynorhinus rafinesquii*). Widespread but uncommon in the se. U.S. Occurs in summer in old cabins in Great Smoky Mts. NP (TN/NC) and in hollow beeches in Natural Bridge SP (KY). The darker-colored southern subspecies is common in summer under road bridges in Kisatchie NF (LA), and in hollow tupelos in swamp forests along the Gulf Coast, for example, in Atchafalaya

Rafinesque's Bat occurs in old cabins and hollow trees in summer.

NWR (LA), Cat Island NWR (LA), and Panther Swamp NWR (MS); it also lives in bat houses in Trinity River NWR (TX). Winters in caves and mines; forages in forests and riparian thickets. Note that big-eared bats tuck in their enormous ears during hibernation, so in winter they can be better identified by their nose shape.

TOWNSEND'S BIG-EARED BAT (*Corynorhinus townsendii*). Common and widespread in w. U.S.; local in BC; very rare in the East. Roosts and winters in caves, mines, and buildings; forages over woodlands, canyons, and overgrown fields. Present year-round in boulder caves in Pinnacles NP (CA), where main roosts are usually closed off, but a few bats are often visible in open parts of caves. In winter look for it in lava tubes in Lava Beds NMT (CA) and in Oregon Caves (OR). In summer try lava tubes of El Malpais NMT (NM), Lewis and Clark Caverns SP (MT), buildings on Santa Cruz I. in Channel Islands NP (CA), and rock niches and crevices in Capitol Reef NP (UT) and City of Rocks NR (ID). Lone males occasionally present from late spring to fall in almost every small cave or abandoned mine in Sierra Nevada foothills (CA). Comes at dusk to drink from the pond at the entrance to Borrego Palm Canyon in Anza-Borrego Desert SP (CA). The pale-colored southern race occurs in small numbers in many TX caves and in deep rock fissures in Wichita Mts. WR (OK). The very rare eastern subspecies (**VIRGINIA BIG-EARED BAT**) occurs in summer in Natural Bridge SP (KY) and in winter at Grandfather Mt. (NC). The large, reddish-colored Ozarks subspecies (**OZARK BIG-EARED BAT**) is even rarer; it lives in caves of n. AR and adjacent OK. Contact Ozark Plateau NWR (OK) about the possibility of accompanying biologists studying it. Both these subspecies are listed as endangered in the U.S.

MEXICAN BIG-EARED BAT (*Corynorhinus mexicanus*). Only recently found in N. America, where it reportedly occurs in Cave Creek Canyon (AZ). Common in caves and abandoned mines in nw. Mexico, particularly at the bottom of Copper Canyon (Chihuahua); feeds in pine-oak forests.

PALLID BAT (*Antrozous pallidus*). Common in desert canyons of w. U.S., for example, in the Grand Canyon (AZ). Roosts in caves, mines, buildings, and rock crevices; forages over desert and scrubland. Can sometimes be found by spotlighting as it moves on the ground, hunting for crickets, scorpions, and small vertebrates. There is a colony in one of the administrative buildings at the Phoenix Zoo (AZ), and another one at the headquarters of White Sands NMT (NM). There are also colonies in abandoned ranch houses in Carrizo Plain NMT (CA); the houses are off-limits, but you might obtain permission to watch the bats emerge at dusk. Individual bats can sometimes be seen in summer in visitor centers and the inn in Petrified Forest NP (AZ), and in swarms of smaller bats at the creek crossing near Wolfe Ranch in Arches NP (UT).

MEXICAN FREE-TAILED BAT (*Tadarida brasiliensis*). Also known as Brazilian Free-tailed Bat, it is the most common and widespread free-tailed bat of N. America. In CA, AZ, NM, and TX it forms enormous colonies in caves and concrete structures. Mass emergences from such colonies can be seen in summer at Carlsbad Caverns (NM), Eckert

Free-tailed bats can't take off from a level surface. If you find a downed bat, put on gloves and try placing it on a tree or a wall it can climb.

James River Bat Cave PR (TX), Frio Cave (TX), Old Tunnel SP (TX), Bat Cave in El Malpais NMT (NM), Congress Avenue Bridge in Austin (TX), Waugh Drive Bridge in Houston (TX), University of Phoenix Stadium (AZ), and both ends of I-80 overpass in Yolo Bypass WLA (CA). The world's largest bat colony, 20 million strong, is in Bracken Cave (TX), where emergence-viewing tours are conducted by Bat Conservation International in summer and early fall (watch for albinos among the emerging bats). Smaller colonies exist in rocky walls of the Grand Canyon (AZ) and Sego Canyon (UT). A colony of a few hundred bats forms in May–September under the pedestrian bridge across Covell Blvd. west of F St. intersection in Davis (CA), where the bats can be seen up close if you don't mind the precipitation (bring a small flashlight!). It is the most common bat species living under bridges in the Sierra Nevada foothills (CA). A small colony is often present in spring inside the basketball stand at the guest parking lot at Archbold Biological Station (FL). These bats also live in bat houses along Museum Rd. on the University of Florida campus in Gainesville (FL) and in Samuel Houston Jones SP (LA). Winters in Mexico and in small numbers in the Southeast; forages high above fields and other open areas.

POCKETED FREE-TAILED BAT (*Nyctinomops femorosaccus*). An abundant tropical species, relatively uncommon in the deserts of sw. U.S. Roosts in rock crevices, caves, and buildings. Small colonies can be found in rock crevices along the Rio Grande in Big Bend NP (TX), in abandoned mines in and around Death Valley NP (CA), and under the roofs of abandoned buildings in Ironwood Forest NMT (AZ). Look for these bats flying at dusk around the observatory at Kitt Peak (AZ).

BIG FREE-TAILED BAT (*Nyctinomops macrotis*). Widespread in arid lowlands and foothills of sw. U.S.; easily identifiable in flight. Small colonies can be found in lava tubes of El Malpais NCA (NM), in rock crevices at Wilderness Ridge (NM), in canyons of Grand Staircase-Escalante NMT (UT), between granite boulders in Joshua Tree NP (CA), and in abandoned houses in the mountains of sw. CO. Large inaccessible colonies exist below Mohave Point and Abyss overlooks on the South Rim in Grand Canyon NP (AZ), where these bats can often be seen flying along the cliffs on summer evenings, and near the waterfall in Pine Canyon in Big Bend NP (TX). Winters in Mexico.

FLORIDA BONNETED BAT (*Eumops floridanus*). Recently (2004) split from Wagner's Bonneted Bat (*E. glaucinus*), a widespread tropical species. Probably endemic to s. FL, where it's rare and local. Roosts under roof tiles and in hollow trees; can often be identified in flight by swiftlike silhouette and loud flight calls. Sometimes seen feeding at dusk over golf courses in the city of Coral Gables and above the trees along Pineland Trail in Everglades NP; but these two locations are not as good as they were a few years ago, and might not be worth trying by the time you read this book. Occasionally comes at dusk to drink from a pond located at 25°30'17" N, 83°23'13" W near Homestead. One individual was present inside the basketball stand at the guest parking lot at Archbold Biological Station in the fall of 2010. A roost exists in a bat house at Babcock Webb WMA, but seeing it requires permission from the management area personnel. Listed as endangered in the U.S.

WESTERN BONNETED BAT (*Eumops perotis*). Sometimes called **WESTERN MASTIFF BAT**, this is the largest N. American bat, with wingspan of more than a foot (30 cm). Uncommon in desert canyons of sw. U.S. It has very loud flight calls and makes a loud swooshing sound with its wings when flying. Roosts in crevices in rock faces and tall buildings. I have seen these bats flying overhead in Anza-Borrego Desert SP (CA), in the Chiricahua Mts. (AZ), and along the Rio Grande in Big Bend NP (TX), but the only colony I could ever find was in Mexico, in an abandoned mine above Naica (Chihuahua). Look also in Capote Canyon (TX) and in Pinnacles NP (CA).

UNDERWOOD'S BONNETED BAT (*Eumops underwoodi*). A tropical species, known in N. America only from s. AZ. You can identify it in flight by its high-pitched calls, so loud that they hurt some people's ears. Roosts in hollow trees, woodpecker holes in large cacti, and dry palm fronds; forages over deserts and woodlands. Occurs at watering holes in Organ Pipe Cactus NMT (AZ).

VELVETY FREE-TAILED BAT (*Molossus molossus*). Also known as **LITTLE** or **PALLAS'S MASTIFF BAT**, this cute tropical species is known in N. America in only the w. Florida Keys. Roosts under roofs made of corrugated iron; forages over towns and woodlands. Look for these small, swallowlike bats flying over city roofs in Marathon, on Boca Chica Key, and in Key West (all in FL). A few related species occur in Mexico and might be eventually found in s. AZ and TX.

CARNIVORES

Carnivores are the most sought-after mammals, but even some common and widespread species can be frustratingly difficult to find. Hunters have developed numerous methods of attracting them, and the tools of their trade are commercially available: scented bait, audiotapes with mating calls of cougars and distress calls of prey species, mouse squeakers, and so on. All carnivores eat carrion whenever available, and they are attracted to meat. Many are also attracted to garbage and regularly visit campgrounds and garbage dumps. If you try baiting, make sure the animals don't learn to associate food with people, for such a habit can be lethal for them.

In the West, many carnivores frequent prairie dog towns and ground squirrel colonies. If the first snow of the season is unusually late, or the snowmelt occurs early, go look for weasels—their white fur will stand out against the dark background. Learning the art of tracking can help in finding many species, especially in the North. Many carnivores regularly patrol their territories and can be expected to revisit the areas where their tracks are present (cats do so about every three weeks or so). Some species (Red Fox, Coyote, Cougar) often rest in places with commanding views of the surroundings; watch for them calmly observing you from haystacks, hilltops, and rocky outcrops.

Most carnivores are predominantly nocturnal, but can often be seen during the day, especially in areas with little or no human disturbance. Spotted and Hog-nosed Skunks, Black-footed Ferret, and Ringtail are almost strictly nocturnal. Small Indian Mongoose, bears, White-nosed Coati, and otters are mostly diurnal, although bears often learn to use roads and garbage dumps late at night, and generally shift to being nocturnal in areas with lots of people. In the Far North, all carnivores can be frequently seen in daylight in summer. If there is a lot of snow (like in the Sierra Nevada and the Cascades in late winter/early spring), many species begin to use rural roads more frequently.

Some species, such as large skunks and small weasels, can be surprisingly tame and easy to follow; they might even approach you and sniff on your toes sometimes if you stand still. Flocks of small birds, especially jays and magpies, often scold small carnivores, and their alarm calls can be a useful clue. Deer alarm calls might alert you to the presence of large predators such as wolves and Cougar.

DOG (*Canis familiaris*). Introduced to N. America by the first human immigrants thousands of years ago, with more recent introductions of different breeds by each subsequent wave of immigration. Feral dogs are surprisingly rare in N. America, except on some tribal lands and in HI. The only truly established wild population, of ancient origin, is the so-called **CAROLINA DOG**. Almost the entire population of this dog has been trapped out for obscure "scientific" purposes (it's difficult not to suspect that the true reason was the prospect of commercial breeding). However, a few Carolina Dogs still survived at the Savannah River Site (SC) at least until 2009; you can search for them by driving late at night along Hwy. 125. Note that Dog is by far the most dangerous mammal in N. America, with millions of attacks on humans (dozens of them lethal) occurring every year. Despite their rarity, feral dogs are responsible for a noticeable share of these attacks. Wild Carolina Dogs are very shy and not known to ever have attacked humans.

GRAY WOLF (*Canis lupus*). Common in much of w. and n. Canada; increasingly common in nw. U.S., MN, MI, WI, and WY, with single individuals dispersing S. to CA, AZ, and CO. The most reliable place to see wolves is Lamar valley in Yellowstone NP (WY), where almost all of them are black (expect traffic jams whenever they show up near a road, particularly in summer). Occasionally seen in Denali NP (AK), in and around Wood Buffalo NP (AB/NT), in and around Jasper NP (AB), around Ross Lake in North Cascades NP (WA), in Voyageurs NP (MN), in Sax-Zim Bog (MN), in Seney and Tamarac NWRs (MN), in Necedah NWR (WI), in Hiawatha NF (MI), along Stewart-Cassiar and Alaska Hwys. (BC/YT), along Dempster Hwy. (YT/NT), along Contwoyto Lake ice road (NT/NU), and along Route du Nord (North Rd.), James Bay Rd., Eastmain Rd., Trans-Taiga Rd., and other roads in n. QC. Reportedly common and easy to see in Thelon WS (NT/NU). Sightings on roads are more common in winter when deep snow forces wolves to use them, but I've also seen wolves on Alaska Hwy. in broad daylight in June despite heavy snowbird motor home traffic at that time. The largest wolves live in n. AB, in sw. NT, and on Kenai Peninsula (AK). Coastal and island wolves of BC and s. AK are genetically distinct; they can sometimes be seen around Yakutat (AK) and along logging roads on Prince of Wales I. (AK). A white race (**ARCTIC WOLF**) occurs on the northernmost islands of NU and in n. Greenland. The smaller Mexican subspecies (**MEXICAN WOLF**), once extinct in the wild, is being reintroduced to Apache-Sitgreaves NFs (AZ) and Blue Range WA (NM). Wolves in the Great Lakes region are hybrids with Algonquin Wolf. Wolves are among the few mammals that are as fun to hear as to see; listen for pack howls just after sunset. Attacks on humans are very rare (just two lethal ones in N. America this century) and mostly involve wolves habituated to people. Listed as endangered in parts of the U.S. where they don't regularly occur; most populations are managed by states and are in danger of falling victim to local politics.

RED WOLF (*Canis rufus*). A 2011 study suggests that "Red Coyote" might be a more appropriate name for this species, which has probably evolved as a result of ancient hybridization between Gray Wolf and Coyote. Once common and widespread in the East, it is now extinct in much of its range. There are two surviving forms, differing in color and the share of wolf versus coyote genetic heritage. The northern race, often considered to be a separate species called **ALGONQUIN WOLF** or **EASTERN WOLF** (*C. lycaon*), occurs mostly in and around Algonquin PP, east to Ottawa suburbs and south to Rideau River PP (all in ON); it is also present in extreme s. QC. It is not particularly shy, and the chances of seeing it in one winter week in Algonquin PP are reasonably good, especially if you drive along the main park road a lot, or if there is a deer carcass placed near the visitor center to attract wolves (that happens most often during the spring school break). The southern race (Red Wolf proper) was exterminated in the wild by the U.S. Fish and Wildlife Service, but has since been reintroduced in and around Alligator River NWR (NC), where it is very difficult to find. The refuge conducts "wolf howls" a few times a year; these are night walks giving you a chance to hear wolves. There is a breeding island population in St. Vincent NWR (FL), but only day visits to the island are currently allowed, so the chances of seeing a wolf there are slim. In 2012 there was a possible sighting in e. LA. Red Wolf (only the southern form) is listed as endangered in the U.S.

COYOTE (*Canis latrans*). Abundant in the West; increasingly common in the East, where it has recently formed a new subspecies (**EASTERN COYOTE**), very distinctive in appearance and carrying some wolf and dog genes. Occurs in all habitats, but prefers open areas and wood margins in the West, and open forests in the East. Coyotes become very tame in some national parks, and can often be seen on roadsides, for example, in Yosemite NP, Carrizo Plain NMT, Joshua Tree NP, Death Valley NP (all in CA), along the access road to the North Rim of the Grand Canyon (AZ), and in Yellowstone NP (WY/MT/ID). Other good places are Cherokee WMA (OK), Parkhill Prairie PR (TX), Antelope Island SP (UT), Fish Springs NWR (UT), and Tranquille WMA (BC). In the North it is most common in the arid interior of e. AK and cen. YT; look for it along Klondike Hwy. (YT). The Eastern form is more difficult to see; good places to try are Three Lakes WMA (FL), Kissimmee Prairie Preserve SP (FL), Hatfield Knob in North Cumber-

Thanks to its intelligence and adaptability, Coyote is now the most commonly seen predator in many parts of the continent.

land WMA (TN), Cades Cove in Great Smoky Mts. NP (TN), Shenan-doah NP (VA), Land Between The Lakes NRA (TN/KY), Cape Breton Highlands NP (NS), and Aiguebelle NP (QC). Urban populations exist in many cities, such as Los Angeles (CA), Boulder (CO), and Baton Rouge (LA), but these animals are often easier to hear than to see. Coyotes in the area from NJ to NS are wolf hybrids, big enough to be dangerous (a hiker was killed by a pack in NS in 2009). Other coyotes (mostly urban) have attacked small children.

RED FOX (*Vulpes vulpes*). Widespread throughout most of N. America, but local in the sw. U.S. and uncommon in the Southeast. Abundant in interior Canada, where it's often seen on roadsides. Most common along forest edges, riparian corridors, and overgrown gullies. In some areas, such as in cen. BC, more than half of foxes are silver-black. One reliable place to see both black and red morphs is at the out-skirts of Churchill (MB). In the U.S., the gorgeous silver-black form can sometimes be seen on winter nights around ski lodges on Mts. Baker and Rainier (WA). Red morph is common, among other places, in Hitchcock NA (IA), Laramie River Greenbelt (WY), Linville Gorge (NC), Great Meadows NWR (MA), and Prince Edward Island NP (PE). Lives in some cities; look for it at Fairmount Cemetery in Denver (CO), in Gateway NRA (NY/NJ), and in all towns of n. QC. In winter, when human access to the base of Niagara Falls is closed, you can some-times watch foxes from the observation tower in Niagara Falls SP (NY), as they search for stunned fish along the water edge just below the falls. The distinctive Sierra Nevada subspecies, which is usually blonde-colored in summer and gray in winter, is extinct everywhere except in and around Lassen Volcanic NP, around Sonora Pass (both in CA), at Mount Hood, and in Crater Lake NP (both in OR). The popu-lation in the Sacramento Valley in CA has been recently shown to rep-resent a previously undescribed native subspecies (**SACRAMENTO VALLEY RED FOX**); look for it in Sacramento NWR. Distinctive races exist also on Kodiak I. (AK) and Newfoundland; both are common throughout their island ranges. The largest foxes occur in the n. Great Plains (particularly in ND and w. MN), as well as on Kenai Peninsula (AK), and in NS. Eastern Red Foxes have been introduced to coastal CA, where they do a lot of damage to local wildlife, particularly around San Francisco Bay.

SWIFT FOX (*Vulpes velox*). Widespread but rare in the shortgrass prairie zone; gets replaced by Red Fox in irrigated places, agricultural areas, and around human dwellings. The best site to look for it is the land of the Blackfeet Tribe (MT). Other places worth trying are Co-manche NG (CO), the western part of Pawnee NG (CO), Badlands NP (SD), Oglala NG (NE), Lake Rita Blanca SP (TX), Rita Blanca WMA (OK), and Hwy. 56/412 between Springer and Clayton (NM). Reintro-duced to Canada, but still very rare and local; try Grasslands NP (SK). Smaller foxes living in arid grasslands of the sw. U.S. are often con-sidered a separate species, **KIT FOX** (*V. macrotis*). Driving across NV

American Black Bear is virtually always black in the East (top), but in the West it is often brown (bottom left), and in some populations you can see beautiful white or dark blue individuals (bottom right).

BC, and Harding Icefield Trail in Kenai Fjords NP (AK). Black bears fishing for salmon can be watched in October along Taylor Creek on the sw. shore of Lake Tahoe (CA). In ON and s. QC, commercial bear-viewing tours, often using feeding stations, are widely available, but these bears are no longer truly wild. The rare, stunningly beautiful blue morph (known as **GLACIER BEAR**) can sometimes be seen around Yakutat (AK); the chances are said to be similar to those of winning a lottery, but I scored in fewer than 16 hours of driving up and down the only road out of town. The white morph (called **KERMODE** or **SPIRIT BEAR**) can be seen in coastal BC south of Prince Rupert, particularly along small coastal creeks during the salmon run. The two largest subspecies occur on Haida Gwaii and Newfoundland; look for the former in Gwaii Haanas NP (BC) and for the latter in Gros Morne NP (NL). The rare FL race can sometimes be found along Big Cypress Bend Trail in Fakahatchee Strand PR SP, on Bear I. in Big Cypress NPR, in Ocala NF, and also in orange groves of cen. FL during fruiting seasons. Another rare subspecies occurs in LA, and can sometimes be seen on Weeks I., in Kisatchie NF, Palmetto Island SP, and Atchafalaya NWR. Black Bears are traditionally considered safe to be around, but they actually kill more people in N. America than Brown Bears (20 versus 16 between 2000 and 2013), so use caution, especially around large males in AK and Canada. The Louisiana subspecies (including bears in sw. MS and e. TX) is listed as threatened in the U.S.

BROWN BEAR [*Ursus arctos*]. The largest terrestrial predator in the Americas, it is still common in parts of nw. N. America, but extinct in the southern part of the range. Prefers open habitats such as tundras, meadows, and clearings, but can also live in dense forests, particularly along the Pacific Coast and around hibernation time. The smaller inland form (**GRIZZLY BEAR**) is common in interior AK and parts of w. Canada, but rare in the U.S., except in Yellowstone NP (WY/MT/ID), where it is most often seen in the northeastern part of the park. Reliable in summer in tundra parts of Denali NP (AK) and around Granite Park Chalet in Glacier NP (MT); in spring along the northern portion of Stewart-Cassiar Hwy. (BC); and in the fall on the outskirts of Deadhorse (AK). In late spring, bears can sometimes be seen feeding on roadkill along Dempster Hwy. (YT/NT). Although generally extinct in the prairie zone, Grizzlies can still be seen in summer on the prairie in Blackleaf WMA (MT) and in Lamar Valley in Yellowstone NP (WY). The larger coastal form (**KODIAK BEAR**) can be seen almost anywhere along the southern coast of AK, where it is abundant around (and sometimes in) Yakutat, and reliably seen on Glacier Bay NP boat tours and on numerous bear-viewing tours to Katmai NP, Kodiak I., and other remote locations. Occurs also in coastal BC around Prince Rupert. A genetically distinct population inhabits Alexander Archipelago (AK). Brown Bear is traditionally considered dangerous in N. America, but attacks on humans are generally rare (usually 1 or 2 per year). Bear spray has been shown to be effective in preventing attacks as long as you know how to use it and carry it at hand rather than in a backpack. Populations in the Lower 48 states are listed as threatened in the U.S.

POLAR BEAR [*Ursus maritimus*]. Widespread but uncommon along the Arctic Coast. Hunts on ice floes and rocky shores; pregnant females den on hill slopes. Bears of the Hudson Bay population spend the summer in coastal tundra; others might soon have to do the same as polar ice caps continue to shrink. Reliable in late spring at Barrow (AK) and in the fall near Churchill (MB); in both places taking a bear-viewing tour is possible, but a rental four-wheel drive will often suffice. Reportedly common in summer around Pond Inlet and Arctic Bay

Volcanic NP, upper parts of Kings Canyon NP, and the southern unit of Giant Sequoia NMT (all in CA). To track martens, choose winter days with no wind and lots of unstable snow on tree branches, when the animal's movement through the trees can be followed by looking for the snow it has dropped on the ground.

STONE MARTEN (*Martes foina*). Also known as **BEECH MARTEN**, (but not having to do much with beech), this Eurasian species has been introduced to the Milwaukee area (WI), where it occurs, among other places, in Kettle Moraine SF. It has broader habitat preferences than American Marten, and can survive in isolated groves, overgrown rocky outcrops, and Christmas tree plantations; some European populations live in cities, feeding on pigeons and rats.

FISHER (*Martes pennanti*). Widespread in boreal forests of Canada; rare but locally recovering in the U.S. More common in the eastern part of the range. Good places to see it are roads in and around Wood Buffalo NP (NT/AB) in summer, and Algonquin PP (ON) in winter (look in the northeastern part of the park and around bird feeders at the main visitor center). Occurs in Hoopa Valley (CA), North Cascades NP (WA), Rydell NWR (MN), Allegany SP (NY), Pisgah SP (NH), Wapack NWR (NH), Esther Currier WMA (NH), Umbagog NWR (NH/ME), Baxter SP (ME), and La Mauricie NP (QC). It has recently colonized Ottawa suburbs; successfully reintroduced in WV. An isolated population occurs in the s. Sierra Nevada, mostly at medium elevations, but seeing Fisher there is very difficult; try Golden Trout Wilderness (CA).

LEAST WEASEL (*Mustela nivalis*). This tiny, very brave hunter is widespread but uncommon in boreal forests; rare in the tundra, the Midwest, and the Appalachians. Prefers open areas with lots of rodents, such as grassy bogs, marshes, and lush meadows, but can also occur in dense forests and city parks. Individual territories are usually very compact and can be located in winter by snow tracking. Often can be attracted by using a mouse squeaker. Common in Algonquin PP (ON), around Chapleau (ON), in Assinica WR (QC), in forest meadows around the Manicouagan Impact Crater (QC), and in Goose Bay area (NL). Occurs in Lonsdale Marsh in Pawtucket (RI). Known locations in the Appalachians include Beauty Spot near Unaka Mt. in Cherokee NF (TN), and Weaver's Bend (TN), located at 36°N, 83°W.

ERMINE (*Mustela erminea*). Also known as **STOAT** or **SHORT-TAILED WEASEL**, this beautiful predator of voles and lemmings is widespread in AK, Canada, Greenland, and colder parts of the U.S. (south to NM and PA), but generally rare in the southern part of its range. Inhabits coniferous forests, tundras, and overgrown fields; often prefers forest clearings and wetlands. Occurs farther north than any other small carnivore; lives in houses in Pond Inlet (NU) and other villages in the Canadian Arctic and Greenland. Guanella Pass (CO) is one of the best places to look for it in the U.S. Reportedly common in Glacier Bay NP (AK), in Lassen Volcanic NP (CA), at Clear Lake (IA), in City of Rocks NR (ID), and in Prince Albert NP (SK), but I've never seen it in any of these places. In the North it is easier to find after "lemming years" (years when lemming populations peak); look for it around Nome (AK), Inuvik (NT), and Longue Pointe (QC). Occasionally seen around Hellroaring Creek Trail parking lot in Yellowstone NP (WY), and in wetlands of Esther Currier WMA (NH). In drought years, when the Sierra Nevada doesn't get the usual amount of snow, look for it in early winter or spring around Taylor Creek Visitor Center at Lake Tahoe (CA).

LONG-TAILED WEASEL (*Mustela frenata*). The commonest and most widespread weasel in the U.S.; rare in s. Canada and local in the Southeast. Can occur in almost any habitat, from desert washes to alpine tundra and city parks, but prefers densely vegetated areas near water. In summer, mostly nocturnal and can be seen on night drives in areas with lots of rodents, such as oak-juniper and piñon woodlands in the West and fallow fields in the East. In the West it often hides under cattle guards installed across rural roads. In winter it tends to be more diurnal; look for it at Point Reyes NSS (CA), in West Kentucky SWA (KY), in Bartlett Arboretum (CT), in Sleeping Bear Dunes NLS (MI), in Lake Ilo NWR (ND), in Ottawa NWR (OH), and on talus slopes above Island Lake in Ruby Mts. (NV). A good place to look in summer is Hidden Lake Overlook Trail in Glacier NP (MT). Occasionally seen in Yellowstone NP (MT/ID/WY), particularly along Mt. Washburn, Lost Lake, and Hellroaring Creek Trails. Try also Owyhee Uplands Back Country Bwy. (ID/OR), City of Rocks NR (ID), and Silver Creek Preserve (ID). "Bridled" southwestern form can be seen, for example, in Bosque del Apache NWR (NM) and in the foothills of Chisos Mts. in Big Bend NP (TX). The beautiful golden form occurring in s. CA is rare and very difficult to find; try Tule Elk SR or Carrizo Plain NMT.

BLACK-FOOTED FERRET (*Mustela nigripes*). Once thought extinct in the wild, this largely subterranean resident of shortgrass prairies can now be found by spotlighting at reintroduction sites, such as in Conata Basin at the southern edge of Badlands NP (SD), the lands of Cheyenne River and Rosebud Sioux Tribes (SD), Shirley Basin (WY), UL Bend NWR (MT), Grasslands NP (SK), and particularly Aubrey Valley (AZ), where the best sites are between mile markers 129 and 125 on Rte. 66 west of Seligman. Search prairie dog towns at night for bright blue or green eyeshine. The Arizona Game & Fish Dept. sometimes conducts ferret-watching tours in Aubrey Valley; call (928) 774-5227 for the schedule. A nonreintroduced population has recently been discovered on the lands of Standing Rock Sioux Tribe (SD), but it is so small and vulnerable that you should avoid visiting it. Listed as endangered in the U.S.

AMERICAN MINK (*Mustela vison*). Widespread in N. America, except for sw. deserts; rare and local in the tundra and in FL. Inhabits wetlands and rocky seashores. Despite being partially diurnal, it can be difficult to see even where common; this is one of those animals that is usually around but seems to disappear whenever you are looking for it. Try canoeing down South Fork Sable River (MI), or walking at dusk along Laramie River Greenbelt (WY), in Pipestone NMT (MN), in Spicer Lake Nature PR (IN), along rivers in Mt. Robson PP (BC), in Sea Rim SP (TX), in Anahuac NWR (TX), in Big Wall Lake NA (IA), in Boatwright WMA's Swan Lake Unit and Ballard WMA (both in KY), in New River SP (NC), in Pillsbury SP (NH), in Umbagog NWR (NH/ME), around beaver ponds in Algonquin PP (ON), in coastal marshes in Prince Edward Island NP (PE); also at low tide on rocky shores in Trustom Pond NWR (RI), around East Ferry (NS), in the coastal part of Kejimkujik NP (NS), in Gros Morne NP (NL), and in San Juan Island NMT (WA). Fish hatcheries throughout N. America and crayfish ponds in sw. LA are also good places to search for it. The rare Everglades subspecies can sometimes be seen at dusk along the eastern side of Shark Valley Loop Rd. in Everglades NP, and north of Tamiami Trail east of Shark Valley turnoff (FL).

AMERICAN BADGER (*Taxidea taxus*). Widespread in the w. U.S.; rare in the Midwest and in sw. Canada. Most commonly seen in shortgrass prairie around prairie dog towns, but can also live in forests, deserts, and mountains. Dens are often located in gullies, ditches, and prairie buttes; it's a good idea to set up an observation blind nearby if you find a den. If found, Badger can be surprisingly easy to approach and observe. Good places to look for it are Point Reyes NSS, Carrizo Plain NMT, Pinnacles NP, and Mineral King Rd. (all in CA); Slough Creek, Lamar Valley, and Soda Butte in Yellowstone NP (WY); Valles Caldera NPR (NM), Comanche NG (CO), Rita Blanca WMA (OK), Aubrey Valley (AZ), Turnbull NWR (WA), Hutton Lake NWR (WY), Thunder Basin NG

Lamar Valley in Yellowstone National Park is one of the best places to see American Badger, especially in winter and spring when there are no crowds.

Northern River Otter is seldom seen far from water; when it is, it's usually because its river or pond has frozen over or dried out.

(WY), Snake River Birds of Prey NCA (ID), Owyhee Uplands Back Country Bwy. (ID/OR), Antelope Island SP (UT), and Badlands NP (SD). Extremely rare in ON; try Port Burwell PP.

NORTHERN RIVER OTTER (*Lontra canadensis*). Widespread over much of N. America, except for arid regions. Extirpated in parts of the U.S., but is now being widely reintroduced. Inhabits permanent bodies of water such as rivers, lakes, and cypress swamps; in the North it depends on openings in the ice for winter survival. Occurs also on rocky seashores, particularly in the Pacific Northwest, BC, New England, and the Maritimes. Places where sightings are most likely include the central part of Yellowstone NP (WY), Soda Butte Creek in Silver Gate (MT), Hawley Wetland in Seedskadee NWR (WY), Paynes Prairie PR SP (FL), Grizzly Island WLA (CA), Cosumnes River PR (CA), Mingo NWR (MO), Hatchie NWR (TN), Great Swamp NWR (NJ), Pillsbury SP (NH), Umbagog NWR (NH/ME), Mystery Trail near Union Village Dam (VT), Muscatatuck NWR (IN), Big Oaks NWR (IN), Sweetwater Strand on Loop Rd. in Big Cypress NPR (FL), interior parts of Okefenokee NWR (GA), Chincoteague NWR (VA), protected coves in Isle Royale NP (MI), Algonquin PP (ON), La Mauricie NP (QC), Fundy NP (NB), and the coasts of Haida Gwaii (BC). A related species, **NEOTROPICAL RIVER OTTER** (*L. longicaudis*), might be present on the Rio Grande in and around Big Bend NP (TX); it has a yellowish snout and its tail is usually longer than 18 in. (45 cm).

SEA OTTER (*Enhydra lutris*). Occurs along the coast of AK from Aleutian Is. to the Juneau area, mostly in protected bays and coves with rocky bottoms and lots of kelp. Reliable places to see Sea Otter include Seward, Homer, Valdez, Yakutat, Kenai Fjords NP, and Glacier Bay NP (all in AK). A small population has recently been discovered on the cen. BC coast. Reintroduced to n. Vancouver I. (BC) and Olympic

in Hole-in-the-Donut area of Everglades NP, along the eastern part of Loop Rd. and along Turner River Rd. in Big Cypress NPR, and in Faka-hatchee Strand PR SP (all in FL). The very rare Eastern subspecies (**EASTERN COUGAR**) has recently been recorded in Forillon NP (QC), in Lake Fausse Pointe SP (LA), and elsewhere in LA, AR, MO, and TN, although some of these animals might be dispersing Cougars from the West. A beautiful sand-colored form, known as **YUMA COUGAR**, occurs along the lower Colorado R., but is very rare; look for it in Pica-cho SRA (CA) and Kofa NWR (AZ). Cougars attack people occasionally, but it happens much less often than the media would make you believe: there have been only three lethal attacks in N. America this century. Florida Panther and Eastern Cougar are listed as endan-gered in the U.S.; the latter also in Canada.

JAGUARUNDI (*Puma yagouaroundi*). A widespread tropical species; inhabits dry tropical forests, dense scrub, and wood margins. There has been no material proof of its presence in N. America since the 1980s, but sightings are regularly reported in extreme s. TX, FL, and recently AL and GA. The FL population is said to be introduced, but there is no good evidence to support this claim. I photographed Jag-uarundi tracks at Archbold Biological Station (FL) in 2007. In TX, try Lower Rio Grande Valley NWR, which still has a few patches of good habitat despite having been largely destroyed by border fence con-struction. Listed as endangered in the U.S.

OCELOT (*Leopardus pardalis*). A beautiful tropical cat, once common north to LA, but extirpated from almost all of its N. American range. A few dozen still live in s. TX (mostly in Laguna Atascosa and Lower Rio Grande Valley NWRs) in dry tropical forests and dense thickets. Virtu-ally extirpated in AZ, where it's been recently recorded only twice, in Huachuca Mts. and Whetstone Mts. As long as the proposed AZ seg-ment of the border fence is not built, recolonization of AZ is still pos-sible, since the nearest breeding population in Mexico is less than 30 miles south of the border. Listed as endangered in the U.S.

CANADA LYNX (*Lynx canadensis*). Widespread in boreal forests of AK and Canada; rare in the Lower 48 states. Common for two to three years every decade in interior AK following peaks of Snowshoe Hare population density. The last time this happened was in 2009–2010. Good places to look for Lynx in those years include Denali NP and the western part of Wrangell-St. Elias NP. In Canada, look for it on sum-mer nights along roads in and around Wood Buffalo NP (NT/AB), in the Chic-Choc Mts. (QC), and around Faro (YT). The Newfoundland subspecies has beautiful silver-gray winter fur; look for it in Gros Morne NP (NL). Very rare in the Lower 48 states; places worth trying include North Cascades NP (WA), Little Pend Oreille NWR (WA), Kan-iksu NF (ID), Superior NF (MN), and Baxter SP (ME). Populations in the Lower 48 states are listed as threatened.

BOBCAT (*Lynx rufus*). Widespread and common in much of the U.S., but rare in Canada and missing from parts of the Great Lakes region. Occurs in all habitats, but mostly in woodlands, rocky areas, riparian corridors, and shrublands. Regularly seen in Point Reyes NSS, in Pinnacles NP, in Joshua Tree NP, along the road to Mineral King in Sequoia NP, along remote stretches of Hwy. 1, and along Nacimiento-Fergusson Rd. (all in CA); along Owyhee Uplands Back Country Bwy. (ID/OR); in Buenos Aires NWR and other protected areas of s. AZ; along Art Project Rd. in Black Bayou Lake NWR (LA); in Highlands Hammock SP and St. Marks NWR (both in FL); along Hwy. 125 through Savannah River Site (SC); in Laguna Atascosa, Santa Ana, and Anahuac NWRs (TX); from the Hatfield Knob Elk Viewing Tower in North Cumberland WMA (TN); in DiLane Plantation WMA (GA); in

With some effort, Bobcat can be seen almost everywhere in the U.S.

Necedah NWR (WI); in the vicinity of Elk Country Visitor Center in Benezette (PA); and in Crab Orchard NWR (IL). A very large, spotted form occurs in the Maritimes and n. New England; it can be found in Cape Breton Highlands NP (NS) and Moosehorn NWR (ME). Southeastern animals tend to be smaller and darker, with a striped pattern, while those from TX westward are often reddish.

DOMESTIC CAT (*Felis catus*). A particularly destructive introduced species; occurs throughout N. America, but almost always within a mile or two of human settlements. Only some populations in HI and on Catalina I. (CA) appear to be self-sustaining. Feral cats number in the millions and have a disastrous impact on many wild mammals, birds, and reptiles.

JAGUAR (*Panthera onca*). Once widespread from CA to LA, this cat is now extremely rare in N. America; probably only one or two animals remain in AZ. Although a breeding population exists in Mexico just 50 to 100 miles south of the border, successful recolonization of the U.S. by jaguars might become impossible if the proposed construction of the border fence takes place. The best place to look for Jaguar in the U.S. is probably the Whetstone Mts. (AZ). Black individuals have been recorded in the U.S. in the past; I once photographed such a jaguar in

Black Jaguars, like this Mexican one, once inhabited the southern U.S. Large black cats are frequently sighted in the Southeast, but these could also be melanistic Cougars or privately introduced Leopards—nobody knows.

nw. Mexico after watching a trail with lots of jaguar tracks from a nearby hollow tree for a few days. Sightings of large black cats in the se. U.S., particularly in LA, MS, AL, TX, FL, and GA, are frequently reported (recently at the rate of hundreds per year), but it is unclear which species is (are?) involved.

SMALL INDIAN MONGOOSE (*Herpestes auropunctatus*). Introduced to Oahu, Maui, and the Big Island (all in HI), where it occurs in virtually all habitats and at all elevations. Check garbage dumps near roadside restaurants on the Kona Coast and the northern coast of Oahu. Usually tame and easy to observe.

PINNIPEDS (SEA LIONS, WALRUS, AND SEALS)

Most pinnipeds can be best seen when they gather at permanent or seasonal locations called haulouts (if they don't breed there), rookeries, or pupping areas. Many of these places can be easily visited, but some of the best ones are very expensive to access. Still, visiting a fur seal rookery, a harp seal pupping area, or a large walrus haulout is a once-in-a-lifetime experience, and one of the best ways for any naturalist to spend money. Four or five species breed at Point Bennett on San Miguel I. (CA), and can be seen there at relatively low cost if you don't mind an adventurous camping trip and a long hike. Coastal rookeries and haulouts can be extremely sensitive to disturbance, especially when newborns are present and can be easily trampled, so it is better to watch them from a concealed lookout or from afar (bring a birding scope!). There are some wonderful exceptions, such as California Sea Lion rookeries in San Francisco and Monterey, where births can be observed from a few feet away.

Pinnipeds are frequently encountered at sea during whale-watching and fishing trips (except those in the se. U.S.), but usually all you see is a seal or walrus face watching you for a few seconds. Some species, however, can be very playful, acrobatic, and fun to watch in the water, particularly Harp Seals, Fur Seals, and Sea Lions. In Monterey Bay you can occasionally observe funny interactions between California Sea Lions and Humpback Whales.

NORTHERN FUR SEAL (*Callorhinus ursinus*). Breeds in summer in enormous rookeries on Pribilof Is. (AK), where seal watching is well organized, and in smaller rookeries on three islands off CA, of which only San Miguel can be visited (check Channel Islands NP website for info on accessing the rookery at Point Bennett). AK animals migrate

Male Northern Fur Seals establish large harems at rookeries.

south in winter. Sometimes seen at sea on Monterey Bay Whale Watch tours (except in summer) and on other pelagic trips off the Pacific Coast; frequently observed in large numbers during ship reposition cruises between CA and BC.

GUADALUPE FUR SEAL (*Arctocephalus townsendi*). Formerly a common species off s. CA, it now breeds mostly in Mexico, with the largest colony on Guadalupe I. Occasionally breeds at the fur seal rookery on San Miguel I. (CA) and on the Farallon Is. (CA). Rarely encountered at sea around Channel Is., and very rarely farther north, to Point Reyes NSS and occasionally n. to AK. The most reliable way to see it is a trip to Guadalupe I. (Baja Norte, Mexico), which fishing and diving tours occasionally visit. Listed as threatened in the U.S.

STELLER'S SEA LION (*Eumetopias jubatus*). Common at many locations along AK and BC coast; for example, in winter in Nanaimo Harbor (BC) and in summer in Kenai Fjords NP and Misty Fjords NMT. Frequently seen from all AK ferries. Used to be abundant in summer on Aleutian Is., but the population there has recently crashed, probably because of climate change. The most unusual and spectacular rookery is located in a huge grotto called Sea Lion Caves (OR). Breeding colonies also exist on islands off CA, of which only San Miguel can

Sea Lion Caves, Oregon.

be visited (Channel Islands NP website has the info on trips to the rookery at Point Bennett). Distant views are possible on Farallon Is. tours from San Francisco. Sometimes seen at sea on Monterey Bay Whale Watch tours (CA). Listed as threatened in the U.S. (Aleutian Is. population as endangered). Animals living e. of Valdez (AK) are slightly smaller and darker.

CALIFORNIA SEA LION (*Zalophus californianus*). Abundant along CA coast; less common farther north. Major breeding colonies are on Channel Is., including San Miguel I. (CA). Smaller rookeries form in spring in Monterey Harbor (where hundreds of nonbreeding animals can be seen year-round) and on many offshore islands in CA and OR. Usually present also at Pier 39 in San Francisco, and in Sea Lion Caves (OR). Watch for these sea lions among much larger Steller's Sea Lions on rocky islets and near the mouths of salmon-rich rivers along WA and BC coasts, particularly in late summer and early fall. You can swim and dive with wild California Sea Lions in La Jolla Cove in La Jolla (CA). During Monterey Bay whale-watching tours, you can often see hundreds of California Sea Lions feeding and playing with dolphins and whales. Recorded north to AK.

WALRUS (*Odobenus rosmarus*). A common species of shallow Arctic seas. The slightly larger Pacific race forms huge coastal haulouts; the only permanent haulouts in the U.S. are on Round I. near Dillingham, which is accessible by expensive prearranged tours, and on Nushagak Peninsula in Togiak NWR, which is practically inaccessible (both in AK). Much larger haulouts form in late summer on the Siberian side of the Bering Strait, but none of them is easy to get to. Recently, late-summer haulouts have been observed on remote shores of nw. AK in years with abnormally low sea ice cover. The Atlantic race mostly hauls out on ice, and can be seen on trips from Pond Inlet (NU) and to Disko Bay (Greenland). It also forms small (up to 50 animals) coastal haulouts, for example, in Northeast Greenland NP in August–September, but all of them are in very remote places. Lone individuals can occasionally be spotted from any coastal location in n. AK, such as Barrow, Nome, and Gambell, and are often seen on boat tours from Upernavik (Greenland).

HARBOR SEAL (*Phoca vitulina*). Common on both coasts, but normally absent from the se. U.S. On the Atlantic Coast, hauls out in winter and early spring at many locations south to NJ, including, for example, Monomoy NWR (MA), the southern end of Salisbury Beach (MA), Sakonnet Point (RI), Montauk Point State Park (NY), Silver Point CP (NY), and Sandy Hook in Gateway NRA (NJ). In summer it is more difficult to see—look for it around Bonaventure I. (QC), where it can sometimes be encountered underwater during scuba diving trips, and in St. Paul's Inlet in Gros Morne NP (NL), where seal-viewing boat tours are available. Abundant year-round on Sable I. (NS), which has

Harbor Seal in Yakutat Bay, Alaska.

the largest breeding aggregation in the Atlantic. A smaller form lives in remote freshwater lakes on Ungava Peninsula (QC). Seals on the Pacific Coast tend to have more variable coat patterns; good places to see them are glacier fronts in Yakutat, LeConte, and Glacier Bays (AK); Kenai Fjords NP (AK); river estuaries around Nome (AK); Kachemak Bay (AK); much of the BC coast; western and northern coasts of the Olympic Peninsula (WA); Strawberry Hill in Neptune SP (OR); Salt Point SP, Point Lobos SR, Moss Landing, and Bolinas Lagoon (all in CA). In Monterey (CA) there is a small haulout south of Marine Science Station; you can also get close-up views at Fisherman's Wharf in the early afternoon, when semi-tame seals wait for arriving fishing boats. A nearby breeding area is in Carmel Bay, where pups can be seen in April and May. Swimming with tame seals is often possible in Children's Cove in La Jolla (CA). Rare in Greenland, where it's listed as endangered.

SPOTTED SEAL (*Phoca largha*). Also known as **LARGHA SEAL**, this is mostly an Asian species, occurring in N. America only in the Bering Sea and along the northern coast of AK and YT. Breeds in late winter on ice floes in the Bering Sea. Large numbers haul out in spring and summer on remnant ice and sandbars at a few remote locations in AK, such as at Kasegaluk Lagoon, Cape Espenberg in Bering Land Bridge NPR, and Kuskokwim Bay. In late summer a few Spotted Seals can be seen from shore around Nome, Barrow, and Gambell (all in AK), but are difficult to distinguish from the more common Harbor Seals unless you can get close. Southern populations are listed as threatened in the U.S.

Spotted Seal breeds on ice, but occasionally can be found on sandy beaches and tidal shallows.

RINGED SEAL (*Phoca hispida*). Despite being regularly hunted by native peoples and Polar Bears, this small, beautifully patterned seal is very common in shallow Arctic waters. It can be seen in winter and spring on ice floes near Barrow (AK), Nome (AK), Churchill (MB), Longue Pointe (QC), and Chisasibi (QC); year-round near Pond Inlet (NU), in Arctic Bay (NU), around Upernavik (Greenland), and in Northeast Greenland NP. A small population lives year-round in Nettilling Lake on Baffin I. (NU). Breeds in snow-covered dens on ice floes, usually on steadfast ice near shore. If newborn pups are exposed because there isn't enough snow, don't approach them: they don't swim well yet and might get in trouble as they try to escape.

GRAY SEAL (*Phoca grypa*). Until recently considered to belong to its own genus and called *Halichoerus grypus*, this seal is very common in the N. Atlantic. Easy to see at many locations from Labrador to Cape Cod, such as the Gulf of St. Lawrence (QC), any rocky islets in the

Maritimes, Acadia NP (ME), the southern end of Salisbury Beach (MA), and Monomoy NWR (MA). The largest breeding colony is on Sable I. (NS); smaller ones form on pack ice in the Gulf of St. Lawrence and on remote islands in New England, such as Green I. (ME), Seal I. (ME), and Muskeget I. (MA). Often seen underwater during scuba diving trips to Bonaventure I. (QC). In winter and early spring it hauls out farther south, and can often be seen at Sakonnet Point (RI) and rarely on Long I. (NY).

HARP SEAL (*Pagophilus groenlandicus*). Abundant in the nw. Atlantic, but seldom occurs close to shore, except in late summer and early fall. This pelagic species is most easily seen by taking a helicopter tour to its pupping areas from Magdalen Is. (Îles de la Madeleine) (QC) in late March. During those tours you can usually see females up close and pet newborn pups; beautifully patterned males are a bit more shy. Another, much larger pupping area is located east of Newfoundland, but only seal hunters visit it. A few pups can reportedly be seen in Sydney Harbor (NS). Huge herds of adult Harp Seals are often seen at sea during Greenland cruises, and occasionally from ferries in NL. Uncommon in late summer around Pond Inlet (NU) and Upernavik (Greenland), and in spring near Tadoussac (QC). In winter and early spring, stray individuals have been recorded south to NC, and show up once every few years at Sakonnet Point (RI) and in Shinnecock Bay (NY) a few hundred yards east of Ponquogue Bridge.

Baby Harp Seals are among the very few wild mammals that can be touched safely (for you and for them). A trip to the ice floes where these seals breed is a unique experience, and you help develop an alternative to the seal hunt.

RIBBON SEAL (*Histriophoca fasciata*). A stunningly beautiful pelagic species; breeds on Bering Sea ice floes in late April. Like the previous species, which also breeds on sea ice outside Polar Bear range, it lacks antipredatory defenses, so adults and particularly pups can be closely approached. The only reliable way to see it is by chartering a boat or a plane for a long trip to its pupping areas (expect a bumpy ride). Hopefully, as mammal watching becomes more popular, someone will eventually organize regular tours from Nome (AK). Occasionally seen on ice floes near Nome in winter, and east to Barrow in spring. Stray individuals are sometimes seen in summer on sand spits in Bering Land Bridge NPR (AK), and have been recorded south to CA.

BEARDED SEAL (*Erignathus barbatus*). This large seal is common in shallow Arctic waters. Can be seen in winter and spring on ice floes near Barrow (AK), Churchill (MB), Longue Pointe (QC), and Chisasibi (QC); year-round near Pond Inlet and Arctic Bay (both in NU), around Upernavik (Greenland), and in Northeast Greenland NP. To get closer views you can use an old Inuit trick: make a white shield and hold it in front of you while crawling up to the seal.

HOODED SEAL (*Cystophora cristata*). A common but difficult-to-see pelagic species of the N. Atlantic. The best way to see it is to take a helicopter tour to Harp Seal pupping grounds from Magdalen Is. (Îles de la Madeleine) (QC) in late March. However, because of extremely short lactation time (just four days), Hooded Seals are present on ice for only a week; besides, they are not very common in this area, so a trip to see them is difficult to plan, and success is far from guaranteed. In some years the weather is too bad for flying throughout that week. Much larger aggregations form east of Newfoundland and in Davis Strait, but there are no seal-viewing tours in those areas. A few pups can reportedly be seen in some years in Sydney Harbor (NS). Lone individuals are occasionally seen in summer around Pond Inlet (NU), during Greenland cruises, and in Northeast Greenland NP as they haul out onto ice floes to molt.

HAWAIIAN MONK SEAL (*Monachus schauinslandi*). Common but declining on uninhabited islands from Nihoa to Midway (HI), most of which are closed to visitors. Unless you can afford to charter a plane or a boat to visit Midway, look for this seal on the beaches of Kauai (HI), where it is uncommon but increasing in numbers. Good locations include Kilauea Point NWR, Makahuena Point, Poipu Beach (try Hotel Hilton area), and small beaches on Na Pali Coast (particularly Ke'e Beach); try also boat trips to Niihua and Lehua Is. These seals often seem tame but are in fact very sensitive to disturbance and should never be approached closely. Listed as endangered in the U.S.

Molting Northern Elephant Seals often cover themselves with sand.

NORTHERN ELEPHANT SEAL (*Mirounga angustirostris*). Increasingly common in CA; in summer occurs at sea north to AK. The largest haulouts (all in CA) are on San Miguel I. in Channel Islands NP, in Año Nuevo SP, at Piedras Blancas Lighthouse, in Morro Bay SP, on Farallon Is., and just off Hwy. 1 north of San Simeon. Small groups haul out on remote islets north to BC; individual seals wander all over the N. Pacific and have reached Asia and HI. The best time to visit the haulouts is in winter, when adult males fight and pups are born. Subadults are usually present in summer (when molting) and fall; in the fall they are joined by yearlings, and in spring also by adult females. Although Elephant Seals usually feed in deep waters far offshore, they are sometimes seen at sea in Misty Fjords NMT (AK), from Aleutian Is. ferries (AK), and on Monterey Bay Whale Watch tours in CA, and can be occasionally encountered underwater while scuba diving at Channel Is. (CA).

CETACEANS
(WHALES, DOLPHINS, AND PORPOISES)

Except for a few species occurring close to shore, cetaceans are expensive animals to see. At most locations you need dozens of whale-watching tours (or, better, your own boat) to see all species regularly occurring there, not to mention rare visitors. Getting underwater views is even more expensive (it is usually possible only in the tropics, where visibility is better). In North America, the best place to spend your money on whale-watching tours is Monterey (CA). Monterey Bay Whale Watch maintains an online database of sightings; I once volunteered with them for a year, and in about a hundred trips I saw 22 species of marine mammals—I don't think you can beat this anywhere else in the world. A more expensive option is to take a cruise. Among the Pacific Coast cruises, summer trips by Princess Cruises and spring/summer ship reposition cruises are the best because they get far offshore where you have a good chance of spotting beaked whales and other rarities.

An interesting phenomenon in recent years is an increasingly common appearance of "friendlies," whales prone to approaching boats and interacting with them. This behavior is observed more often at particular locations, and seems to be somewhat contagious. Meeting a friendly is always an unforgettable experience, but you have to be very careful to avoid propeller trauma to the animal. Feeding friendlies should never be attempted.

Most cetaceans are surprisingly easy to identify at sea if you have a few whale-watching trips' worth of experience, but beaked whales can be tricky, so if you see them on a guided tour, don't expect trip leaders or tour guides to be of much help. Almost all *Mesoplodon* sightings on those tours are recorded as "*Mesoplodon* sp." It doesn't mean they are not identifiable at all; you just have to know what features to look for.

Pods of dolphins can often be spotted from afar by watching for seabird flocks circling above; this works particularly well for *Stenella* species. If you are planning a whale-watching trip, be sure to check the marine forecast, and try to pick a day with the calmest conditions, not just to avoid seasickness, but also because some species (Harbor Porpoise, small beaked whales, Northern Minke Whale, and particularly Pygmy and Dwarf Sperm Whales) are virtually impossible to spot in rough seas. (A good idea is to take your first Dramamine pill about six hours before the trip, and the second one about an hour before departure.)

Almost all cetaceans occasionally get stranded onshore; some species are more prone to this than others. If you live in a coastal area, consider joining the local marine mammal rescue organization (if there is one); they are often based in local universities. Volunteering as a rescuer gives you a good chance to see and help save a few rarities, as well as more common species.

Note that cetacean systematics are outdated and might change substantially in the near future. I expect some genera to be eventually lumped, and a few species to be split.

GRAY WHALE (*Eschrichtius robustus*). Common in the Pacific from AK to CA, where it conducts long annual migrations. This whale is a bottom-feeder and prefers shallow waters. Reliably seen in winter and spring (with peak numbers in December and March) on Monterey Bay Whale Watch tours (CA), and in spring on whale-watching tours from Vancouver I. (BC). Regularly seen from numerous vantage points along the Pacific Coast during fall and particularly spring migration; Point Reyes Lighthouse and Cabrillo NMT (both in CA) are the best places. In summer, frequently seen from shore and ferries in the Aleutians, and from shore on Bering Sea islands; some individuals reach Barrow (AK) and even YT waters. My favorite place to see Gray Whale is Laguna San Ignacio (Baja Sur, Mexico), where in February and March mothers with calves approach small boats to play with people and have their snouts scratched. If you'd like to see mothers with newborn calves in the U.S., try Monterey Bay in April and May, as they migrate north later than other Gray Whales. Recently two stray individuals have been recorded in the Atlantic for the first time since this species was extirpated there centuries ago.

NORTHERN MINKE WHALE (*Balaenoptera acutorostrata*). Common and widespread north to the Chukchi Sea and Davis Strait. Inconspicuous, shy, and difficult to see off the Pacific Coast, but relatively tame and sometimes friendly in New England, e. Canada, and the Arctic. Regularly seen in summer on whale-watching tours from Victoria (BC), Cape Cod (MA), Montauk Point (NY), Grand Manan I. (NB), and Percé (QC). Abundant in summer and fall in the Gulf of St. Lawrence and reliably seen on whale-watching tours from Baie-Sainte-Catherine and Tadoussac (both in QC); sometimes the whales are present in harbors there and can be viewed from shore. Viewing from sea kayak is often possible in June and early July in Bonne Bay in Gros Morne NP (NL). Occasionally seen in summer from AK ferries and various points on the AK coast, as well as on Kenai Fjords NP boat tours (AK) and from San Juan Is. ferries (WA). A small population occurs year-round in and around Monterey Bay (CA), where these whales are occasionally seen on Monterey Bay Whale Watch tours; try also looking from Point Pinos in Pacific Grove (CA) in summer and fall whenever the sea is relatively calm. In spring and fall, they are occasionally spotted during ship reposition cruises between CA and BC. In winter, present off FL and HI, but rarely seen there.

BRYDE'S WHALE (*Balaenoptera brydei*). A warm-water species, very rare in N. America. Recorded year-round off HI and Bermuda, in summer in the Gulf of Mexico and around Channel Is. (CA), and at least once in spring from the FL–Bahamas ferry. Reportedly common off n. Cuba, and might become easier to see if ferries from FL to Cuba ever run again. A small, genetically distinct population has been discovered in 2014 off the FL Panhandle.

SEI WHALE (*Balaenoptera borealis*). Widespread but rare and unpredictable in N. American waters. Seen occasionally on pelagic bird-watching tours off NC, during whale-watching tours off MA and NS, during cruises around s. Greenland, from Aleutian Is. and Haida Gwaii ferries, and very rarely elsewhere. The nearest place where it can be reliably seen (in summer) is a remote area in Denmark Strait between Greenland and Iceland. Listed as endangered in the U.S.

FIN WHALE (*Balaenoptera physalus*). Common in Atlantic and Pacific Oceans, but highly nomadic. Seen in summer on whale-watching tours from East Ferry (NS), Grand Manan I. (NB), Montauk Point (NY), and Cape Cod (MA), as well as in Disko Bay (Greenland). Rarely observed in summer and fall on Monterey Bay Whale Watch tours (CA), and in summer from AK ferries. Regularly seen in summer and fall during ship reposition cruises between CA and BC. Viewing from sea kayak is sometimes possible in June and early July in Bonne Bay in Gros Morne NP (NL). Listed as endangered in the U.S.

BLUE WHALE (*Balaenoptera musculus*). This light-colored giant with a geyserlike spout is rare but increasing in the Atlantic and Pacific Oceans. Seen reliably (except if the water temperature is too high) in summer and early fall on Monterey Bay Whale Watch tours (CA) and regularly on summer whale-watching tours from Percé (QC), as well as on summer tours to the Farallon Is. (CA). A small population is present in summer in the Gulf of St. Lawrence, where these whales are occasionally seen on whale-watching tours from Baie-Sainte-Catherine (QC), and from the Rimouski–Forrestville ferry (QC). Underwater viewing is sometimes possible in summer in the Gulf of California (Mexico). Listed as endangered in the U.S. and Canada.

HUMPBACK WHALE (*Megaptera novaeangliae*). Widespread and increasingly common in N. American waters. The most often seen whale on summer and fall whale-watching tours almost anywhere in N. America: in AK, all along the Pacific Coast, from Cape Cod (MA), Grand Manan I. (NB), Montauk Point (NY), Percé (QC), and East Ferry (NS). Often seen from shore from various points in NS in summer, and in Pacific Rim NP (BC) in spring and fall. Reliable in May through November on Monterey Bay Whale Watch tours (CA), where huge aggregations sometimes occur in late summer and friendlies are increasingly common in recent years. Other reliable options include Farallon Is. tours (CA), Kenai Fjords NP tours (AK), and Glacier Bay NP tours (AK), where sleeping whales are sometimes observed. Viewing from sea kayak is often possible in June and early July in Bonne Bay in Gros Morne NP (NL). Common in summer around Greenland. Very common in winter and early spring off HI, where breaching whales are often seen from planes landing or taking off at Kailua-Kona Airport, and numerous humpback-watching tours are available on all major islands. Abundant and easy to see in early spring around Bermuda. This whale is very popular among whale watchers thanks to the broad variety of spectacular behaviors it often displays. The unique bubble fishing behavior can be observed with some effort from chartered boats in s. AK, especially in Misty Fjords NMT, and rarely in the Bay of Fundy (NB/NS) and in Monterey Bay (CA). Breaching is observed much more often in wintering areas such as HI, but recently also off CA. Humpbacks produce loud, very complex songs that can be heard with a hydrophone or simply by putting your head underwater. Diving with Humpbacks can be arranged in late winter in Turks and Caicos in the Caribbean. Listed as endangered in the U.S.; the Pacific population is listed as threatened in Canada.

NORTHERN RIGHT WHALE (*Eubalaena glacialis*). Rare in the N. Atlantic, where it's regularly seen in late summer on whale-watching tours from East Ferry (NS) and Grand Manan I. (NB); in early May on whale-watching tours from Gloucester (MA); rarely around Cape Cod (MA); sometimes also from ferries to NS, Grand Manan, and Deer I. (NB). Winters and breeds off GA and FL, and is occasionally seen from October through April from FL coast between Cape Canaveral and St. Augustine. The ne. Pacific population, often considered to belong to a different species, **NORTH PACIFIC RIGHT WHALE** (*E. japonica*), is very close to extinction; these whales mostly live far offshore, although there is a very small chance of seeing one from Aleutian Is.

Northern Right Whales are rare, but once found, they can be very easy to observe.

ferries and off Kodiak I. (AK). Listed as endangered in the U.S., Canada, and Greenland.

BOWHEAD WHALE (*Balaena mysticetus*). An Arctic species, usually occurring along the edge of ice floes. Regularly seen from shore in Barrow (AK) in late spring; it is also possible to arrange a viewing trip on an Inupiat whaling boat from there. Often seen in summer around Igloolik (NU), near Arctic Bay (NU), in Disko Bay (Greenland), and on Greenland cruises. In winter it occurs in the Bering Sea and off Labrador. Listed as endangered in the U.S., Canada, and Greenland.

BELUGA (*Delphinapterus leucas*). Common and widespread in the Arctic, rare and local farther south. Reliable in summer at Beluga Point near Anchorage (AK), in Churchill (MB), and around Pond Inlet (NU), near Arctic Bay (NU), and in Disko Bay (Greenland); in the fall near Qaanaaq (Greenland). A small population lives in the Gulf of St. Lawrence, where sightings are possible from all ferries and sometimes from shore at Rivière-du-Loup and Tadoussac (both in QC). Often seen in Mackenzie R. near Inuvik (NT) in late summer. The best views (even underwater) are possible in summer near Anadyr Airport (Russia). Listed as endangered in Greenland, the U.S. (Cook Inlet population), and Canada (populations in Gulf of St. Lawrence and parts of Hudson Bay).

NARWHAL (*Monodon monoceros*). This most unusual-looking whale is common in the Canadian Arctic and locally around Greenland. Reportedly reliable on summer tours from Pond Inlet (NU), to Karrat Fjord (Greenland), and in Disko Bay (Greenland); also in spring near Qaanaaq (Greenland). Up to 100 gather in summer in Scoresbysund Fjords in Northeast Greenland NP. Arctic Kingdom runs very expensive Narwhal-viewing tours to Arctic Bay (NU), where snorkeling with Narwhals is possible.

ROUGH-TOOTHED DOLPHIN (*Steno bredanensis*). Widespread in warm waters; generally tame and will sometimes bow-ride, but very difficult to find in N. America. Sometimes seen on dolphin-viewing tours in HI, especially around Kauai I. Occurs in the Gulf of Mexico and around Bermuda. Was once present north to VA, but seems to have disappeared from the Atlantic Coast. Recorded off CA.

COMMON BOTTLENOSE DOLPHIN (*Tursiops truncatus*). Widespread in tropical and temperate waters, often very close to shore, in bays, harbors, and estuaries. In the Atlantic, distinctive inshore and offshore populations exist and might eventually be recognized as separate species. The Atlantic inshore form is the most likely dolphin to be seen from shore in the Gulf of Mexico and along the Atlantic Coast north to Cape Cod, occasionally to NS. Reliable in early spring in

Flamingo (FL) near the harbor entrance. Frequently seen year-round at many locations along the Atlantic Coast, for example, from Port Aransas and Galveston I. ferries (TX), Cumberland Island NSS (GA), Fort Macon SP (NC), and Kiptopeke SP (VA). Dolphin-viewing tours are offered by Florida Aquarium in Tampa (FL), by a few operators in the Florida Keys, and also from Hilton Head I. (SC). The dark-colored Atlantic offshore form can occasionally be seen on pelagic bird-watching tours from Hatteras (NC), from Key West–Dry Tortugas ferries (FL), on diving trips to Flower Garden Banks NMS (TX), around Bermuda, and in summer from NS ferries. Offshore and inshore forms also exist in the Pacific; they both differ from the ones in the Atlantic in having a dark stripe between the eye and the beak, but look very much alike. The Pacific inshore form can often be seen from shore in CA, particularly from Moss Landing and Torrey Pines SR. The Pacific offshore form is often seen on dolphin-viewing tours from Oahu (HI). It also occurs very far off CA, often in mixed herds with Short-finned Pilot Whale.

PANTROPICAL SPOTTED DOLPHIN (*Stenella attenuata*). Widespread in tropical waters; recorded north to Cape Cod (MA) in summer. Occasionally seen on dolphin-viewing tours from HI, from shore in Padre Island NSS (TX), from FL–Bahamas ferries, on diving trips to Flower Garden Banks NMS (TX), and on fishing trips from Florida Keys and Bermuda.

ATLANTIC SPOTTED DOLPHIN (*Stenella frontalis*). Occurs along the Atlantic Coast, north to NY in summer; particularly common in the Gulf of Mexico. Occasionally seen in summer on pelagic bird-watching tours from Hatteras (NC), and year-round from Ft. Myers–Key West (FL) and Florida–Bahamas ferries. Occurs in Flower Garden Banks NMS (TX) and around Bermuda. Swimming with wild dolphins of this species is offered by a few tour operators in the Bahamas. Spotted dolphins living in the waters south of NS and Newfoundland possibly represent a separate subspecies; they are occasionally seen from ferries there.

CLYMENE DOLPHIN (*Stenella clymene*). Also known as **SHORT-SNOUTED SPINNER DOLPHIN**, this Atlantic species occurs in the Gulf of Mexico year-round, and north to NJ in summer, but usually far offshore. I've seen it during a diving trip to Flower Garden Banks NMS (TX) in early spring. Like many other tropical cetaceans, it can be frustratingly difficult to find; pods of dolphins are often separated by hundreds of miles of "empty" ocean.

SPINNER DOLPHIN (*Stenella longirostris*). Known for its acrobatic leaps, this widespread tropical species is common in HI and reliably seen on dolphin-viewing tours there; occasionally visible from

In warm-water years, immense herds of long-beaked common dolphins move up the California coast as far as Point Arena.

Kilauea Point NWR. Pacific animals have a habit of resting in protected bays during the day; the ones in the Atlantic tend to stay far offshore. Frequently recorded in summer and early fall in the Gulf of Mexico; rare off the Atlantic Coast north to NC and around Bermuda.

STRIPED DOLPHIN (*Stenella coeruleoalba*). A widespread warm-water species, uncommon in N. American waters. Occurs off the Atlantic Coast (mostly off FL, but occasionally north to NS and even s. Greenland in summer) and rarely off HI and Bermuda. Recorded off s. CA and very rarely north to BC, but always far offshore. Look for it from FL–Bahamas ferries, and during diving trips to Flower Garden Banks NMS (TX).

LONG-BEAKED COMMON DOLPHIN (*Delphinus capensis*). Common and widespread in warm coastal waters; in N. America it occurs only off CA mostly in late summer and early fall. Often seen in immense herds on Monterey Bay Whale Watch tours (CA), and regularly during boat crossings to Channel Is. (CA); also occasionally on pelagic birdwatching tours from central CA.

SHORT-BEAKED COMMON DOLPHIN (*Delphinus delphis*). Common and widespread in warm offshore waters, north to OR, Cape Cod (MA), and occasionally to NL in summer; rare in winter. Occasionally seen from August to October in very large herds on Monterey Bay Whale Watch tours (CA), and more regularly in summer and early fall on boat crossings to Channel Is. (CA); also on pelagic bird-watching tours from NC, whale-watching tours from Montauk Point (NY) and Grand Manan I. (NB), and rarely from FL–Bahamas ferries.

FRASER'S DOLPHIN (*Lagenodelphis hosei*). A widespread tropical species, beautifully patterned, but very little known and rarely seen. Occurs off HI, often in mixed herds with other dolphins and small whales. Strandings recorded in the Gulf of Mexico, near Miami (FL), and in Bermuda, but there seem to be no records of live dolphins ever being observed there.

ATLANTIC WHITE-SIDED DOLPHIN (*Lagenorhynchus acutus*). Common from Cape Cod (MA) to NL; occurs also around s. Greenland in summer, and south to DE in winter. Regularly seen on whale-watching tours from Percé (QC) in July and August, from Cape Breton I. (NS) from July through October, from Bar Harbor in Acadia NP (ME) in August through October, and from Gloucester (MA) in May and October. Sometimes visible from NL and NS ferries, and from shore on Bonaventure I. (QC).

WHITE-BEAKED DOLPHIN (*Lagenorhynchus albirostris*). Occurs in summer from Cape Cod (MA) to s. Greenland. Regularly seen from late July till late September on whale-watching trips from St. Anthony (NL) and Cape Breton I. (NS). Reportedly seen sometimes from various ferries in NL, and occasionally from Confederation Bridge connecting NB and PE. In winter it has been recorded from Nantucket ferries (MA), but very rarely.

PACIFIC WHITE-SIDED DOLPHIN (*Lagenorhynchus obliquidens*). Common (mostly in summer in the North) from the Gulf of Alaska to CA; the most often seen dolphin in much of offshore N. Pacific. Large herds are regularly seen year-round (most often in February through April and less often in summer) on Monterey Bay Whale Watch tours (CA), commonly in summer and fall around Vancouver I., and occasionally in summer around the Aleutian Is. (AK). Often forms mixed groups with Risso's and Northern Right Whale Dolphins.

NORTHERN RIGHT WHALE DOLPHIN (*Lissodelphis borealis*). This beautiful little dolphin in desperate need of a better common name is common off the Pacific Coast, north to Vancouver I. and sometimes to the Aleutian Is. (AK). Usually occurs far offshore, but large herds are frequently seen (particularly in summer and early fall) on Monterey Bay Whale Watch tours (CA).

RISSO'S DOLPHIN (*Grampus griseus*). Widespread and common, but mostly far offshore. Recorded north to AK and NL in summer. Regularly seen year-round, but more often in winter, on Monterey Bay Whale Watch tours (CA), often in huge mixed herds with Pacific White-sided and Northern Right Whale Dolphins. Occasionally encountered on summer pelagic bird-watching tours from Hatteras (NC). Recorded in the Gulf of Mexico, off HI, and around Bermuda.

LONG-FINNED PILOT WHALE (*Globicephala melas*). Common in the N. Atlantic, where it is sometimes seen in summer on whale-watching tours, and occasionally from shore or sea kayak, especially along the ME coast from Casco Bay to Machias Bay, in Bonne Bay in Gros Morne NP (NL), and in Terra Nova NP (NL). Reliably seen on whale-watching trips from the northern tip of Cape Breton I. (NS) from July through October, often with dolphins or Northern Minke Whales. Pilot whales are highly prone to mass strandings, which frequently occur from NL to NC.

SHORT-FINNED PILOT WHALE (*Globicephala macrorhynchus*). A widespread warm-water species. Occurs along the Atlantic and Gulf Coasts from TX to NC, and around Bermuda; isolated records farther north. Sometimes seen on whale-watching tours off NC, and rarely in the waters of Everglades NP (FL) and around the Florida Keys. In the Pacific it is regularly recorded around Channel Is. and very far off cen. CA, as well as on dolphin-viewing tours around HI.

MELON-HEADED WHALE (*Peponocephala electra*). An obscure tropical species, seldom seen close to shore. Occurs far off FL; sometimes seen off HI and can potentially be seen on ship cruises from FL to Bermuda. Isolated records north to NC and around Channel Is. (CA), all from July through October. Reportedly common around Palmyra Atoll in Remote Pacific Islands MNM.

PYGMY KILLER WHALE (*Feresa attenuata*). A widespread but uncommon and little-known tropical species. Whales from the small resident population in HI are occasionally seen on dolphin- and whale-watching tours. Recorded off FL and around Bermuda; look for it in the warm waters of the Gulf Stream.

Killer Whales are believed by many people to be the world's most beautiful mammals.

FALSE KILLER WHALE (*Pseudorca crassidens*). A widespread tropical species, rare in N. America. Usually found over deep water. Herds sometimes appear off CA during warm-water years. Occasionally seen on dolphin- and whale-watching tours from HI and on bird-watching tours from Hatteras (NC). Isolated records north to NY and AK, and from the Gulf of Mexico. Populations appear to be rapidly declining because of accidental killings by longline fishery.

KILLER WHALE (*Orcinus orca*). Also known as **ORCA**, this stunningly beautiful predator is very widespread and locally common. Numerous "ethnic groups" (traditionally called "types") differ in appearance, language, and culture, and are probably going to be split into an uncertain number of species in the near future. Three such types occur along the Pacific Coast. **"RESIDENT" KILLER WHALES**, which feed mostly on schooling fish like salmon and herring, occur in a string of isolated populations in inshore waters. They can be reliably seen on Orca-viewing tours in the Seattle area (WA), and are frequently seen on Kenai Fjords NP and Glacier Bay NP boat tours (AK) and from AK ferries. Recently some have wintered in Monterey Bay (CA), probably due to lack of salmon up north. The Puget Sound population appears to be on its way to extinction due to a lack of salmon and constant disturbance by whale-watching boats; it is listed as endangered in the U.S., but the conservative Pacific Legal Foundation is working hard to make sure this population is stripped of legal protection and goes extinct. Highly nomadic **"TRANSIENT" KILLER WHALES** (also known as **BIGG'S KILLER WHALES**), which hunt marine mammals, are most often seen on Monterey Bay Whale Watch tours (CA), especially during Gray Whale migration in late January through May, and also in September through November when they mostly hunt California Sea Lions. Friendlies have appeared there in recent years. Killer Whales of this type are also sometimes seen from Aleutian Is. ferries (AK) in summer. It is possible that there are also small groups of Killer Whales off CA that specialize in killing either Sperm Whales or great white sharks, but they travel widely, and too little is known about their habits. **"OFFSHORE" KILLER WHALES**, which live in large groups and hunt squid and fish such as sharks, salmon, and tuna, are only occasionally seen on Monterey Bay Whale Watch tours, and on overnight pelagic bird-watching tours from elsewhere in CA (mostly in fall and winter), but are often spotted during ship reposition cruises between CA and BC. In the Atlantic, Killer Whales are much more difficult to see. They are occasionally spotted on whale-watching and pelagic bird-watching tours anywhere from SC to NU, as well as on Greenland cruises and from NS and NL ferries.

HARBOR PORPOISE (*Phocoena phocoena*). This tiny porpoise is locally common in temperate coastal waters, from AK to CA and from s. Greenland to SC, but is inconspicuous and easy to overlook, particularly in choppy seas. Recent acoustic studies have also found it over deep waters of cen. N. Atlantic. Regularly seen in summer from Vancouver I. ferries (BC), Kenai Fjords NP tours (AK), and Monterey Bay Whale Watch tours (CA), as well as in Forillon NP (QC) and around Tadoussac (QC). Often seen in spring and summer from Golden Gate Bridge (CA), usually near the northern end at high tide, and in late summer in Yakutat Bay (AK). At the southern tip of Deer I. (NB) it can sometimes be seen feeding in the world's second-largest whirlpool (better from a kayak). Pacific animals belong to a darker subspecies. The Atlantic population is listed as threatened in Canada.

DALL'S PORPOISE (*Phocoena dalli*). Fast and showy, this porpoise is common in small groups along the Pacific Coast, from Aleutian Is. (AK) to CA. Frequently seen year-round (but mostly in winter and spring) on Monterey Bay Whale Watch tours (CA), year-round on whale-watching tours from Victoria (BC), and in spring through fall in Telegraph Cove on Vancouver I. (BC). In summer, often seen from AK ferries, as well as on Misty Fjords NMT and Kenai Fjords NP tours (AK). In spring and summer, reliably seen during ship reposition cruises between CA and BC.

BAIRD'S BEAKED WHALE (*Berardius bairdi*). Locally common in the N. Pacific over deep water. Occurs year-round (but mostly in the fall) off CA, where small groups are sometimes seen on Monterey Bay Whale Watch tours, Farallon Is. tours, and pelagic bird-watching tours from Bodega Bay. Present in summer around Aleutian Is. (AK) and off Vancouver I. (BC). In some years common in June and July around Guadalupe I. south of the U.S./Mexico border.

NORTHERN BOTTLENOSE WHALE (*Hyperoodon ampullatus*). Locally common over deep water in the N. Atlantic. Occurs far offshore from s. Greenland to NC. There are two areas where it is regularly present, but both are very difficult to access: off n. Labrador (NL), and in The Gully MPA (NS). Once found, it can be friendlier and easier to observe than other bottlenose whales. Very rarely seen from ferries in NL and NS.

CUVIER'S BEAKED WHALE (*Ziphius cavirostris*). Also known as **GOOSEBEAK WHALE**, this widespread species is more tolerant of shallow water than other beaked whales, so it is regularly recorded on both coasts, north to AK and NS. Common far offshore around FL; regularly occurs west of San Clemente I. (CA). Occasionally seen on pelagic bird-watching tours off NC and CA, for example, above Bodega Submarine Canyon, and on dolphin-viewing tours off HI. Occurs in the Gulf of Mexico; very common around Guadalupe I. south of the U.S./Mexico border.

SOWERBY'S BEAKED WHALE (*Mesoplodon bidens*). A rare, little-known Atlantic species. Most sightings in N. American waters occur over deep submarine canyons at the edge of the continental shelf, particularly in The Gully MPA (NS) and at the head of McMaster Submarine Canyon (39°44'N, 71°39'W); the latter location can be reached by chartering a fishing boat from e. Long I. (NY). Strandings are most frequent in Scotland and Norway, but also from NL to Cape Cod (MA), so there is a small chance of seeing it from NL and NS ferries. One record from the w. coast of FL.

GERVAIS'S BEAKED WHALE (*Mesoplodon europaeus*). The most common *Mesoplodon* whale off the Atlantic Coast of the U.S. and in the Gulf of Mexico,. Recorded from TX to ME, but most strandings occur in FL and NC. Look for it on pelagic bird-watching tours off NC. I've seen it in early spring in Flower Garden Banks NMS (TX).

TRUE'S BEAKED WHALE (*Mesoplodon mirus*). Known in N. America from strandings on the Atlantic Coast (from FL to NL). It is suspected that this small whale is associated with the Gulf Stream, so it is theoretically possible to see it on pelagic bird-watching tours off NC, from FL–Bahamas ferries, and from cruise ships going to Bermuda. Like all beaked whales, it spends up to 80 percent of its time in long, very deep dives, and so is difficult to find.

BLAINVILLE'S BEAKED WHALE (*Mesoplodon densirostris*). A widespread deep-water species with almost totally unknown natural history. Although it has been recorded north to WA and NL, as well as in the Gulf of Mexico, it seems to be very rare everywhere along both coasts, except the area from FL to MA, where strandings are more frequent, and the deep waters east of the Bahamas. Look for it on pelagic bird-watching tours off NC. There are also a few sight records south of Grand Isle (LA). Acoustic studies suggest that it is extremely rare off the Pacific Coast, but somewhat more common off HI, where it is occasionally observed, mostly off the Waianae Coast of Oahu.

STEJNEGER'S BEAKED WHALE (*Mesoplodon stejnegeri*). The most common *Mesoplodon* whale of the N. Pacific and the Bering Sea. Unlike many other beaked whales, it can sometimes be seen close to shore as long as the water is deep. Relatively common in summer in the waters south of Commander Is. (Russia), less so off w. Aleutian Is. Acoustic studies suggest that it is also common far off WA. Regularly recorded south to CA.

HUBBS'S BEAKED WHALE (*Mesoplodon carlhubbsi*). A little-known whale of the deeper parts of the N. Pacific; acoustic studies have shown that it is very rare. Some experts believe that this whale occurs across the N. Pacific; others think it is limited to areas where cold currents enter warm waters. Males are easy to identify at sea if seen at close range, but this whale is very shy and difficult to approach. Sightings have occurred far off OR and s. CA. I saw two whales that I think belonged to this species in Cordell Bank NMS (CA) in August 1999, but, as usual, they disappeared before I could get a really good look at them. Strandings happen once every few years in CA and occasionally north to BC, as well as in HI and Japan. A few individuals have accidentally died in fishing nets. The best option for looking for this species might be to arrange a ride on one of the numerous cargo ships sailing from Oakland (CA) to China, Japan, or S. Korea.

PERRIN'S BEAKED WHALE (*Mesoplodon perrini*). A recent (2002) split from Hector's Beaked Whale (*M. hectori*), a Southern Hemisphere species. Known from a handful of strandings and sight records in CA north to Monterey Bay. This shy whale is believed to be extremely rare, but it is possible that its main area of occurrence lies very far offshore and is yet to be discovered. Acoustic studies suggest that it is not uncommon off s. CA above depths of 3,600 to 4,200 ft. (1,100–1,300 m). I saw it once in May 2003, 150 nautical miles west-south-west of Santa Cruz (CA).

GINKGO-TOOTHED BEAKED WHALE (*Mesoplodon ginkgodens*). A rare, mysterious warm-water species of the Pacific and Indian

Perrin's Beaked Whale is one of many enigmatic cetaceans that occur far from land and are rarely seen; any observation of them is of scientific value.

Oceans. Strandings and sightings (only by whalers so far) are most frequent in Japan; a couple of strandings have happened in CA. Acoustic studies have failed to record it along the Pacific Coast, but suggest that it regularly occurs off HI (particularly around Kauai) and is very common around Cross Seamount south of the Big Island.

PYGMY BEAKED WHALE (*Mesoplodon peruvianus*). A warm-water species of the e. Pacific Ocean, known in N. America from just two strandings in CA (at Moss Landing and at Newport Beach). Acoustic studies suggest that it doesn't normally occur that far north, but is relatively common in the southern part of the Gulf of California (Mexico). Another accidental visitor is **TROPICAL BEAKED WHALE** (*M. pacificus*), a widespread but little-known tropical species (for such a rare animal, it has a surprising variety of alternative names, including **LONGMAN'S BEAKED WHALE** and **TROPICAL BOTTLENOSE WHALE**; it's also often classified in its own genus, *Indopacetus*). A single stranding has recently been recorded in HI; acoustic studies suggest that it normally occurs farther south, for example, around Palmyra Atoll in Remote Pacific Is. MNM.

PYGMY SPERM WHALE (*Kogia breviceps*). Widespread but very rarely seen and little known. Recorded on both coasts north to NL and BC, usually more than 100 miles from shore. Strandings are most frequent in FL, HI, and on Sable I. (NS). Difficult to spot and identify at sea; mimics sharks in appearance. The only time I've seen it was by a lucky chance: it was suspended in a huge wave that rose above the fishing boat I was in.

Sperm Whale can be easily recognized from afar by its spout, which is shot at an angle rather than straight up.

DWARF SPERM WHALE (*Kogia sima*). This tiny, somewhat sharklike whale is believed to be widespread along the edge of the continental shelf. Strandings are most frequent in FL. Occasionally observed far offshore in the Gulf of Mexico and off NC. Recorded off HI and along the Pacific Coast north to BC; genetic data suggest that Pacific animals might represent a separate species.

SPERM WHALE (*Physeter macrocephalus*). The world's largest predator, it is widespread in deep waters, but cannot be reliably found anywhere in N. America (try Martinique or the Azores instead). Males are recorded north to AK and Greenland in summer, while females reach only CA and NC. Single males are seen once every few months on Monterey Bay Whale Watch tours (CA), once or twice per summer on pelagic bird-watching tours from Hatteras (NC), a few times every spring around Bermuda, and rarely year-round off HI. Frequently present in summer in The Gully MPA (NS). A small resident population seems to exist in the e. Gulf of Mexico. Listed as endangered in the U.S.

UNGULATES (HOOFED ANIMALS)

Most North American ungulate species are delightfully easy to see. Some occur at unnaturally high densities because of insufficient predation, since many carnivores have been hunted out. This overabundance creates serious problems for numerous plants and animals. As if this was not enough, many exotic ungulates have been released throughout the southern part of the continent, but mostly in Texas.

Forest ungulates tend to be crepuscular or nocturnal, but can often be seen in the open on cool mornings, particularly in areas with no hunting. Mountain species often go to great lengths to obtain salt, and will sometimes lick cars coming from cities in winter. Ask local hunters or park/forest rangers for locations of salt licks. In summer many ungulates can be seen in the mountains above the timberline, where vegetation is particularly nutritious and biting insects are less numerous. Most species used to migrate, sometimes in immense herds and over long distances, until much of North America was privatized and covered with fences. Nowadays only Caribou can still migrate more or less freely. A few herds of deer, Elk, Pronghorn, and American Bison migrate shorter distances, but suffer horrendous losses from traffic collisions, barbed wire entanglements, and (in the case of Bison) politically motivated culls. There are proposals to restore Bison migrations on the Great Plains, and even an ongoing attempt to do so in American Prairie Reserve (MT), but this is unlikely to happen in our lifetime.

Most newborn ungulates can follow their mothers within a few minutes or hours of their birth, but some forest species such as deer need to stay in hiding for a few days. If you find a tiny deer fawn, don't touch it: mothers have been known to abandon their fawns if they smelled of human hands.

HORSE (*Equus caballus*). Feral mustangs are widespread in the West, but decreasing in numbers. Good places to see them include Bighorn Canyon NRA (MT), Piute Mts. (CA), Wild Horse Island SP (MT), Theodore Roosevelt NP (ND), and Pryor Mt. Wild Mustang Center (WY). A small herd regularly visits Amargosa Opera House in Death Valley Junction (NV) to drink. In the East, managed feral herds exist on Sable I. (NS), Assateague I. (MD), Chincoteague I. (VA), Shackleford Banks (NC), Cumberland Island NSS (GA), and in Paynes Prairie PR SP (FL).

DONKEY (*Equus asinus*). Feral Donkey is less common than feral Horse, and is usually found in more arid places, mostly in the Southwest, where it's known as **BURRO**. Wild Donkeys in NV look very similar to the extinct Nubian subspecies of African Wild Ass (*E. africanus*), their wild ancestor. They can be seen in Marietta Wild Burro Range (NV), Imperial NWR (AZ), Alamo Lake SP (AZ), Mojave NPR (CA), Custer SP (SD), and on the slopes above the Kona Coast (HI).

FERAL PIG (*Sus scrofa*). N. American feral pigs are a mix of introduced European Boars (of at least two subspecies) and escaped domestic pigs. They are heavily hunted and usually very shy, but can be rather noisy at night. Look for them at night in Cherokee NF (TN), Three Lakes WMA (FL), Merritt Island NWR (FL), Lower Suwannee NWR (FL), Bayou Sauvage NWR (LA), Chicot SP (LA), and Aransas NWR (TX). Feral Pigs of CA are mostly of domestic origin and can be very bold (occasionally to the point of attacking people, particularly small children). They are easy to see, for example, in Joseph D. Grant CP, Pinnacles NP, and on Santa Cruz I. Feral Pigs in HI are a mix of Polynesian and European breeds; they can often be seen along the access trail to Puu Oo Crater on the Big Island.

COLLARED PECCARY (*Pecari tajacu*). Once almost extirpated in N. America, this charming, intelligent, and highly adaptable animal is rapidly recolonizing the sw. U.S., where it is known as **JAVELINA**. Inhabits desert scrub, woodlands, and forested canyons. Visits campgrounds at night in Big Bend NP and Falcon SP (both in TX); common in Black Gap WMA, Davis Mts. SP, Choke Canyon SP, and Bentsen-Rio Grande Valley SP (all in TX), as well as in Roosevelt Lake WLA (AZ), Chiricahua NMT (AZ), and around Quemado (NM).

ELK (*Cervus canadensis*). Common in the Rockies and in the Cascades; local elsewhere. Inhabits forests, meadows, and prairies; often grazes above the timberline in summer. There are four surviving races, very different in appearance. **ROCKY MOUNTAIN ELK** can be easily seen in Valles Caldera NPR (NM), Rocky Mt. NP (CO), Grand Teton NP (WY), Yellowstone NP (WY/MT/ID), Silver Creek Preserve (ID), Kootenai NWR (ID), Turnbull NWR (WA), Banff and Jasper NPs (AB), along Hwy. 50 east of Ely (NV), and along the Alaska Hwy. around mile 5,006 (km 1,526) (YT). In winter, huge herds congregate in National Elk Refuge (WY) and along Hwy. 285 in Rio Grande del Norte NMT (NM). Urban populations exist in w. AB. **ROOSEVELT ELK**, a bright-colored northwestern race, is common in Redwood NP (CA), King Range NCA (CA), Dean Creek Elk Viewing Area on Hwy. 38 east of Reedsport (OR), St. Helens NMT (WA), Mt. Rainier NP (WA), Olympic NP (WA), and on Vancouver I. (BC). **TULE ELK**, the smallest race, occurs in a few reserves in s. Central Valley of CA, including Merced NWR, Tule Elk SR, Grizzly Island WLA, and Carrizo Plain NMT (it is a bit more difficult to find in the latter place); there is also an introduced population at Point Reyes NSS (CA). The largest race, **MANITOBA ELK**, can be seen in Elk Island NP (AB), Prince Albert NP (SK), Riding Mt. NP (MB), and Pembina Hills (ND). It has been introduced to a few places within the former range of the extinct Eastern Elk, which was very similar, if not identical. Introduction sites include the southern side of Great Smoky Mts. NP (NC), and North Cumberland WMA (TN), where an observation tower at Hatfield Knob allows good views during rut (late September through early November). There are scattered introduced Elk herds at other locations in the East, for example, on Cape Cod (MA) and around Benezette (PA), where the watchtower at Elk Country Visitor Center allows excellent views during the rut.

Rocky Mountain, Manitoban, and Roosevelt Elk differ in coat color and antler size.

FALLOW DEER (*Cervus dama*), **AXIS DEER** (*C. axis*), **SIKA DEER** (*C. nippon*), and **SAMBAR DEER** (*C. unicolor*) are Eurasian species that have been introduced to numerous locations throughout the U.S., mostly for hunting. All four, particularly Axis Deer, are common in cen. and s. TX, where most live on private ranches and are essentially domestic animals. Fallow Deer is also common, for example, in the inland portion of Point Reyes NSS (CA), but that herd is about to die out as all animals have been sterilized. It also occurs in GA, where it causes a lot of forest damage. Axis Deer has also been introduced to the island of Molokai (HI). Sika Deer herds exist in Chincoteague NWR (VA), while Sambar Deer occurs in St. Vincent NWR (FL).

MULE DEER (*Odocoileus hemionus*). Abundant in the West in habitats ranging from desert shrub to coniferous forests and (in summer) alpine meadows. Occurs in many cities, such as Berkeley (CA) and Victoria (BC). Good places to see various races in scenic environment include Cuyamaca Rancho SP (CA), Mineral King (CA), Big Bend NP (TX), Buffalo Lake NWR (TX), Canyonlands NP (UT), Antelope Island SP (UT), Heart Butte Reservoir area (ND), Medicine Creek SWA (NE), Mt. St. Helens NMT (WA), and Jasper NP (AB). Northwestern races are collectively known as **BLACK-TAILED DEER**; they are easy to see in Glacier Bay NP (AK), Redwood NP (CA), Trinity Alps (CA), and particularly at Hurricane Ridge in Olympic NP (WA) in summer. The largest subspecies (**KAIBAB DEER**) is common from May to October along the access road to the North Rim of the Grand Canyon (AZ). Huge winter concentrations of Mule Deer can be seen in Wellington Deer Range along Upper Colony Road just north of Wellington (NV), in Boise River WMA (ID), and at Oatman Flat north of Silver Lake (OR). Mass migration can be observed in spring and fall from Hwy. 30 east of Sage Junction (WY). Albinos are often seen near Ozette Lake in Olympic NP (WA) and in Gray Lodge WLA (CA). Introduced to Kauai (HI).

WHITE-TAILED DEER (*Odocoileus virginianus*). Abundant in and around deciduous woodlands in the e. U.S., the Midwest, and Canada east of the Rockies. Largely confined to riparian forests in the prairie zone; very local farther west, where it can be seen in National Bison Range (MT), Red Rock Lakes NWR (MT), Peace River Valley (AB), and (mostly in winter) in Little Pend Oreille NWR (WA). The largest animals occur in Wood Buffalo NP (NT/AB). Good places to see newborn fawns in June include Cades Cove in Great Smoky Mts. NP (TN), Big Meadows in Shenandoah NP (VA), Land Between The Lakes NRA (TN/KY), Sachuest Point NWR (RI), and Red Top Mt. SP (GA). Large winter aggregations form along the eastern border of Algonquin PP (ON), in Aiguebelle NP (QC), in Dawson WMA (ND), in Wenlock WMA (VT), and in Maquam Bog in Missisquoi NWR (VT). The dwarf Florida Keys race (**KEY DEER**) can be reliably seen by slowly driving around Key Deer NWR (FL) at dusk or at night. Other distinctive races include **COUES'S DEER**, which inhabits dry pine-oak forests of the Southwest and can

If you find a newborn fawn, don't touch it or pick it up.

be seen in Madera, Ramsey, and Carr Canyons (AZ) and in Fort Bayard (NM); **CARMEN MOUNTAIN DEER**, which is common at higher elevations in Big Bend NP (TX); **COLUMBIA DEER**, which occurs in Julia Butler Hansen NWR (OR); and big-hooved **MCILHENNY'S DEER**, which inhabits the swamps of coastal LA, including Atchafalaya NWR and Lake Fausse Pointe SP. Key and Columbia Deer are listed as endangered in the U.S.

MOOSE (*Alces alces*). Abundant in AK, common but declining in much of Canada and parts of New England; local in the Rockies. Prefers wetlands with lots of willows amidst dense boreal forests, but frequently wanders into open tundra (mostly in summer), meadows, and even prairies; in summer months it can spend many hours grazing chest-deep in small lakes. Easy to see, for example, along the western approach to Cameron Pass (CO), at Owl Ridge Overlook in Arap-

Moose spend a lot of time in the water in summer to avoid biting insects, chill out, and reach tasty aquatic plants.

aho NWR (CO), in Turnbull NWR (WA), in Isle Royale NP (MI), in Prince Albert NP (SK), in Riding Mt. NP (MB), in Algonquin PP (ON), in Aiguebelle NP (QC), in Chic-Choc Mts. (QC), along Trans-Labrador Hwy. (QC/NL), along Robert Campbell Hwy. (YT), in Peace River Valley (AB), in Wells Gray PP (BC), and in city parks in Anchorage (AK). Look for it also in Yellowstone NP (WY/MT/ID), in Grand Teton NP (WY), along the Green River in Seedskadee NWR (WY), at Lake Pend Oreille (ID), in Silver Creek Preserve (ID), in Kootenai NWR (ID), in the Pembina Hills (ND), in the western part of Rocky Mts. NP (CO), and in Sax-Zim Bog (MN). Red Rock Lakes NWR (MT) is a great place to watch moose rut in early October. In New England it might be a bit trickier; look in late spring and summer along Rte. 114 between East Burke and Canaan (VT), in Connecticut Lakes SF (NH), along Rte. 16 south of Errol (NH), and in the moose-viewing area on Rte. 26 ten miles west of Errol. The largest animals occur in interior AK; look for them around Fairbanks and in the eastern part of Denali NP. Tundra-living Moose can be seen in Thelon WS (NT/NU) and along Dempster Hwy. (YT). Introduced to Newfoundland, where abundant in Gros Morne NP.

CARIBOU (*Rangifer tarandus*). Widespread in AK, Canada, and Greenland, but usually occurs far from human-populated areas. Very rare and local in the Lower 48 states. Inhabits forests and tundras with a thick growth of lichens on the ground. The largest race, **WOODLAND CARIBOU**, can be seen in winter on the coast north of Berry Hill in Gros Morne NP (NL), in Jasper NP (AB), in Prince Albert NP (SK), and at various points along Alaska Hwy., for example, around mile 864 (km 1,392) in YT. In summer, look for it at Pink Mt. summit (BC), along Hwy. 40 to Grande Cache (AB), on mountaintops in Gaspésie NP (QC), and on Long Range Traverse Trail in Gros Morne NP (NL). In spring and fall try Eastmain, Chisasibi, Longue Pointe, and Trans-Taiga Roads (QC). It also occurs on Kenai Peninsula (AK), where it can often be seen along Marathon Rd. in winter, and along Bridge Access Rd. through Kenai River Flats near the town of Kenai in August and September. The Alaska race (**PORCUPINE CARIBOU**) can be seen in summer in the western part of Denali NP and in the town of Deadhorse; in early fall along Denali Hwy., along Glenn Hwy. near Eureka, along Richardson Hwy. near Paxson Lake, and along the Alaska Hwy. east from Tok; also at any time of year along Dempster Hwy. (YT/NT) and Dalton Hwy. (AK). Seeing large migrating herds is sometimes possible along these roads, but if you want to witness really spectacular migration, it is better to arrange a trip to Kobuk Valley NP or Noatak NPR (both in AK). The long-antlered Canadian tundra subspecies (**BARREN GROUND CARIBOU**) can be seen around Pond Inlet, Kugluktuk, and many other communities in NU, around Inuvik (NT), and in many places along the western coast of Greenland. Huge migrating herds of this subspecies are often the highlight of a trip to Thelon WS (NT/NU). The small, beautiful, white-coated **PEARY CARIBOU** lives only on the northernmost islands of NU and in far nw. Greenland, and can be seen in Ellesmere Island NP (NU). In Eurasia this species is known as **REINDEER**; introduced Reindeer from Siberia occur as

Migrating Porcupine Caribou are a common sight in remote parts of Alaska.

domesticated or semi-wild animals around Nome and in a few other locations in AK. Woodland Caribou populations in ID and WA are listed as endangered in the U.S.; its populations in QC are listed as threatened in Canada, while Peary Caribou is listed as endangered there.

PRONGHORN (*Antilocapra americana*). Often called **ANTELOPE** in the U.S., this last survivor of an ancient American family is common in western grasslands and sagebrush deserts. Abundant almost everywhere in WY, except for densely forested areas and high mountains. Other reliable places to see it include Grasslands NP (SK), Petrified Forest NP (AZ), Bryce Canyon NP (UT), Antelope Island SP (UT), Badlands and Wind Cave NPs (SD), Clayton Lake SP (NM), Lucerne Valley Campground in Flaming Gorge NRA (UT/WY), northern parts of Big Bend NP (TX), and all NGs in the shortgrass prairie zone. Look for it in summer in Lamar Valley in Yellowstone NP (WY), and in winter north of Gardiner (MT) just outside the park. Within a few decades, huge migrating herds might reappear in American Prairie Reserve (MT). For now, the longest Pronghorn migration happens in ID between Craters of the Moon NMT and Caribou-Targhee NF. The slightly different western race is common in Hart Mt. NWR (OR); in extreme ne. CA, in Carrizo Plain NMT (CA), where it's been reintroduced; and in Sheldon NWR (NV). The rare Sonoran subspecies can be found with some effort in Buenos Aires NWR (AZ); it is listed as endangered in the U.S.

NILGAI (*Boselaphus tragocamelus*) and **BLACKBUCK** (*Antilope cervicapra*) are exotic antelopes introduced to TX, where they are already more numerous than in their native range in India. Nilgai is particularly abundant in s. TX (for example, in Lower Rio Grande Valley NWR); there are also herds in AL. Blackbuck is common in cen. TX; there is also a small herd in the airport area on Catalina I. (CA). Yet another exotic ungulate, **GEMSBOK ORYX** (*Oryx gazella*) from s. Africa, is rapidly spreading in e.-cen. NM, where it causes a lot of damage to fragile desert vegetation.

MOUNTAIN GOAT (*Oreamnos americanus*). Famous for its unbelievable rock-climbing skills, this showy relative of Asian mountain antelopes is locally common in rocky areas with steep slopes above the timberline from AK to ID. It does not always migrate to lower elevations in winter, but sometimes descends into the forest to access salt licks, and can be seen at licks along Icefields Parkway and at Disaster Point in Jasper NP (AB), and past mile 4 (km 8) on Atlin Rd. (YT). Reliably seen in summer along Harding Icefield Trail in Kenai Fjords NP (AK), at Our Lake (MT), during Glacier Bay tours (AK), in Mt. Rainier NP (WA), in Farragut SP (ID), and around Seton Lake (BC). In spring, look for it in Flume Creek Mountain Goat Viewing Area near Metaline (WA). Present year-round in Blackleaf WMA (MT) and Lost Creek SP (MT). Introduced to many locations outside its natural range; of those,

Seeing Mountain Goat used to be extremely difficult, but now a few roads and easy trails reach its montane habitat.

reliable at Baronette Peak in ne. Yellowstone NP (WY/MT) year-round; on Mt. Evans (CO), in Olympic NP (WA), and above Deep Lake in Beartooth Mts. (WY) in summer; and in Clarks Fork Canyon (WY) in winter.

MUSKOX (*Ovibos moschatus*). The largest surviving mammal of the open tundra, it occurs locally in nw. NU (for example, on Devon I., around Kugluktuk, and in Ellesmere Island NP); in far n. NT, particularly in and around Thelon WS (NT/NU); and in parts of Greenland, where large herds live in Northeast Greenland NP and in Ittoqqortoormiit area. Reintroduced herds can be seen around Nome (AK), where they usually hang out on Anvil Mt., but can also be on any of the nearby hills; also along the northernmost part of Dalton Hwy. (AK), and around Kangerlussuaq Airport (Greenland).

BIGHORN SHEEP (*Ovis canadensis*). Locally common in mountains and canyons of the West, usually near steep slopes. The large, stocky Rocky Mt. subspecies (**ROCKY MOUNTAIN BIGHORN SHEEP**) can be easily seen in winter around Jasper (AB), at Radium Hot Springs in Kootenay NP (BC), in National Elk Refuge (WY), in Sun River Canyon (MT), at the bighorn viewpoint on I-70 near Vail (CO), and in Flume Creek Mountain Goat Viewing Area near Metaline (WA); in summer on Mt. Evans (CO) and at Vaseux Lake (BC); year-round, with some effort, in the Tower area of Yellowstone NP (WY), in Rocky Mt. NP (CO), at Antelope Island SP (UT), and in Lost Creek SP (MT). This race has also

Muskox is the last survivor of the so-called mammoth fauna in North America.

been introduced to Badlands NP (SD), where another subspecies, now extinct, used to occur. The small, slender desert subspecies (**DESERT BIGHORN SHEEP**) is more difficult to see, but looking for it often involves hiking through some of the world's most striking landscapes. Places where it can be found regularly include the entrance to Borrego Palm Canyon in Anza-Borrego Desert SP (CA), Barker Dam area in Joshua Tree NP (CA), the main entrance to Valley of Fire SP (NV), Afton Canyon (CA), and Steens Mt. (OR) in summer; the lower slopes of the Grand Canyon (AZ) year-round; and Kofa NWR (AZ) in winter. The website of Sheldon NWR (NV) has a detailed list of locations within the refuge where a sighting is possible. I've also seen Desert Bighorns in Arches and Canyonlands NPs (UT), in Desert NWR (NV), and at Mt. Graham (AZ). They are often seen on boat tours of Canyon Lake (AZ). The rare Sierra Nevada subspecies (**SIERRA NEVADA BIGHORN SHEEP**) can sometimes be seen in summer around Mono Pass in Yosemite NP (CA) and at Rae Lakes in Sequoia NP (CA). This subspecies and the population of Desert Bighorn Sheep from sw. CA are listed as endangered in the U.S.

DALL SHEEP (*Ovis dalli*). Very local in mountains with steep slopes from AK to n. BC. Possibly conspecific with Bighorn and Snow (*O. nivicola*, a Siberian species) Sheep. The white northern form can be reli-

ably seen in summer in Denali NP (AK), in winter along the road from Anchorage to Whittier (AK), and year-round near Atugin Pass on Dalton Hwy. (AK), on Sheep Mt. (AK), and in Kluane NP (YT). The dark southern form (**STONE SHEEP**) can often be seen along Alaska Hwy. in Stone Mt. and Muncho Lake PPs (BC), and at mile 822 (km 1,323) in YT, as well as from South Canol Rd. (YT) at mile 132 (km 213). The rare intermediate form (**FANNIN SHEEP**) occurs along Blind Creek Rd. near Faro (YT), mostly from early fall to early summer.

Dall Sheep are usually present in summer at Polychrome Pass in Denali National Park.

AMERICAN PIKA (*Ochotona princeps*). Locally common at high elevations in the West. Look for it in summer at the summit of Mt. Evans (CO), at Spanish Peaks (CO), along Frozen Lake Trail in Mt. Rainier NP (WA), at Wheeler Peak (NM), around Lake Marie on Snowy Range Bwy. (WY), along Lost Lake Trail or at Hellroaring Trailhead in Yellowstone NP (WY), around the upper cable car station above Teton Village near Grand Teton NP (WY), in higher portions of Lassen Volcanic NP (CA), at a talus slope about three miles up Mono Pass Trail in Yosemite NP (CA), and at Our Lake (MT). Occurs at lower elevations in NV, but is rare and declining there because of climate change; the best places to find it in NV are Great Basin NP and Angel Lake near Wells. The steel-gray northern race can be seen at mile 61 (km 99) of Icefields Pkwy. in Jasper NP (AB). The golden-colored northwestern race inhabits Coquihalla Canyon PP (BC). A dark form occurs in Craters of the Moon NMT (ID), for example, around the amphitheater near campsite 39.

PYGMY RABBIT (*Brachylagus idahoensis*). Widespread but uncommon in sagebrush deserts of the Great Basin. Listen for its alarm call, a rapid sequence of five to eight high-pitched buzzes. The best places to look for it are the Mono Lake area (CA), Big Sheep Creek near Dell (MT), and Fossil Butte NMT (WY). Often seen on night drives along Hwy. 50 and Hwy. 6 (NV), and sometimes in and around Bodie SHP (CA). Other places to try are Bannack SP near Dillon (MT), Craters of the Moon NMT (ID), City of Rocks NR (ID), Bruneau Meadows along Hwy. 747 about 20 miles south of Jarbidge (NV), Hart Mt. NWR (OR), and Owyhee Uplands Back Country Bwy. (ID/OR). Northern populations are listed as endangered in the U.S.

BRUSH RABBIT (*Sylvilagus bachmani*). One of the most common mammals at low elevations in much of CA; uncommon in OR, where it is hybridizing with introduced Eastern Cottontail. Look for it along the edges of brushy patches, for example, in Mt. Diablo SP (CA), in Great Valley Grasslands SP (CA), or in Illinois River Forks SP (OR). Common in many cities, such as Berkeley and Monterey (both in CA). The distinctive riparian subspecies is listed as endangered in the U.S.; it occurs in Caswell Memorial SP and has been reintroduced to Salinas River NWR (both in CA).

EASTERN COTTONTAIL (*Sylvilagus floridanus*). In the U.S., common in the East, the Midwest, and parts of the Southwest; local but increasing in se. Canada; introduced to the Pacific Northwest. Inhabits a wide variety of habitats, but is most abundant along wood margins, meadow edges, and in forest clearings. Easy to see, for example, in Sunset Crater Volcano NMT and Aubrey Valley (both in AZ), in Cedar Hill and Atlanta SPs (TX), in DeSoto NWR (IA), in Tunica Hills WMA (LA), in Platt Branch WMA (FL), around Archbold Biological Station

Eastern Cottontails watching the sunset in Sachuest Point National Wildlife Refuge, Rhode Island.

(FL), in Hoosier NF (IN), in Sachuest Point NWR (RI), in Spicer Lake Nature PR (IN), as well as in the suburbs of Ottawa (ON), Boston (MA), Washington, DC, Atlanta (GA), and Vancouver (BC).

DAVIS MOUNTAINS COTTONTAIL (*Sylvilagus robustus*). A recently (1998) described species, rare and local. Occurs in Davis and Chisos Mts. (TX). Might still occur in Guadalupe Mts. (TX/NM). Look for it in shady, densely overgrown ravines around the McDonald Observatory in Davis Mts. In Big Bend NP (TX) it occurs around Chisos Basin Campground, along the road to Chisos Basin Lodge, and also along Emory Peak, Lost Mine, and Laguna Meadows Trails.

MANZANO MOUNTAINS COTTONTAIL (*Sylvilagus cognatus*). A little-known species recently split from Eastern Cottontail. Occurs in clearings in tall coniferous forest above Fourth of July Campground and along the access road to the town of Manzano in Manzano Mts. (NM). Rabbits occasionally seen along the upper section of the road to Sandia Crest (NM) likely also belong to this species.

NEW ENGLAND COTTONTAIL (*Sylvilagus transitionalis*). Very local in open woodland and along field margins in lowland areas from s. ME to CT. Look for it in the Salt Meadow Unit of Stewart B. McKinney NWR (CT), in the vicinity of Charlestown (RI), and in Mashpee NWR on sw.

Cape Cod (MA). In the latter area, pine barrens along Great Hay Rd., Degrass Rd., and Surf Dr. are said to be particularly good. A southern form with a different number of chromosomes, described as **APPA-LACHIAN COTTONTAIL** (*S. obscurus*), occurs in the s. Appalachians, mostly above 1,640 ft. (500 m). Look for it at Grandfather, Roan, and Mitchell Mts. (NC), at Dolly Sods in Monongahela NF (WV), along the road to Clingmans Dome in Great Smoky Mts. NP (TN/NC), at Beauty Spot Lookout (36°08'N, 82°20'W) near Unaka Mt. in Cherokee NF (TN), along Skyline Dr. in Shenandoah NP (VA), and at Spruce Knob (WV). Yet another distinctive form, possibly an undescribed species, occurs on Cumberland Plateau; it can often be seen along the access road to Hatfield Knob in North Cumberland WMA (TN).

DESERT COTTONTAIL (*Sylvilagus audubonii*). Very common in much of the interior w. U.S. in deserts, shortgrass prairies, and arid wood-lands. Good places to look for it include Pinnacles NP, Mono Lake, Carrizo Plain NMT, Joshua Tree NP, and Salton Sea NWR (all in CA), Scotts Bluff NMT (NE), Badlands NP (SD), and almost any dry brushy area in NV, UT, CO, NM, AZ, WY, and w. TX.

MOUNTAIN COTTONTAIL (*Sylvilagus nuttallii*). Abundant in the inte-rior West, from s. AB to cen.-w. NM. Lives in rocky upland deserts, along woodland streams, and on vegetated mountain slopes. Easy to find in Mono Lake area (CA), Magdalena Mts. (NM), and in almost any open habitat with brushy cover or wood patches in CO, UT, NV, WY, MT, and the Great Basin. Rare in Canada, where it can be found in Okanagan Valley (BC) and Cypress Hills PP (AB/SK).

SWAMP RABBIT (*Sylvilagus aquaticus*). Common in floodplain forests and swamps along the western part of the Gulf Coast and in the lower Mississippi Basin. Good places to find it include Anahuac NWR (TX), Bayou Sauvage NWR (LA), Little River NWR (OK), Sequoyah NWR (OK), Caddo Lake SP (TX), Richland Creek WMA (TX), West Kentucky SWA (KY), Chickasaw NWR (TN), St. Catherine Creek NWR (MS), and Bond Swamp NWR (GA). Sometimes occurs in city suburbs, for ex-ample, in Baton Rouge (LA) and Greenville (MS).

MARSH RABBIT (*Sylvilagus palustris*). Common in wet lowland habi-tats from AL to Great Dismal Swamp NWR (NC/VA). Reliable places to find it include Croatan NF (NC), Savannah NWR (SC/GA), Fort Pulaski NMT (GA), First Landing (formerly Seashore) SP (VA), Bond Swamp NWR (GA), Bill Baggs Cape Florida SP (FL), Three Lakes WMA (FL), Kissimmee Prairie Preserve SP (FL), and Lake Woodruff NWR (FL). Occasionally seen at night along Flamingo Rd. in Everglades NP (FL). Two dwarf subspecies occur in the Florida Keys; they can be seen by slowly driving at dusk along the back roads of n. Key Largo and Key Deer NWR (FL). The latter race is listed as endangered in the U.S.

Marsh (top) and Swamp Rabbits are the world's only lagomorphs adapted to semiaquatic life.

EUROPEAN RABBIT (*Oryctolagus cuniculus*) is an introduced species. It occurs in isolated colonies, mostly on predator-free islands and in city parks, for example, on Anacapa I. (CA), in San Juan Islands NMT (WA), in Valdez (AK), in Victoria (BC), in Anchorage PP on Grand Manan I. (NB), and on Lehua I. (HI). It can be extremely destructive: on the island of Laysan (HI) it caused extinctions of three endemic bird species and almost all endemic plants.

EUROPEAN HARE (*Lepus europaeus*), also known as **BROWN HARE**, is another introduced species, formerly common in agricultural areas of the ne. U.S. and se. Canada. Now it is declining and remains mostly in s. ON, for example, north of Toronto, in fields and meadows.

WHITE-TAILED JACKRABBIT (*Lepus townsendii*). Widespread but uncommon in the interior West. Rare in the Midwest and CA despite broad habitat preferences, from sagebrush deserts to open forests. Look for it in Grasslands NP (SK), in Thunder Basin NG (WY), in Fossil Butte NMT (WY), near Pawnee Buttes (CO), around Waunita Hot Springs (CO), in Rockport SP (UT), along Red Rock Rd. just e. of Red Rocks Lakes NWR (MT), in Willow River SP (WI), in City of Rocks NR (ID), along Owyhee Uplands Back Country Bwy. (ID/OR), and in Hells Gate SP (ID). The rare western subspecies can sometimes be seen in Deep Springs Valley (CA). This species has brilliant red eyeshine, much brighter than the more common Black-tailed Jackrabbit.

ANTELOPE JACKRABBIT (*Lepus alleni*). Locally common in desert grasslands of s. AZ, for example, in Cabeza Prieta NWR, and along Madera Canyon Rd. southeast of Green Valley. Occurs in Organ Pipe Cactus NMT and Saguaro NP (also AZ). Much more common in Mexico, where it can be easily seen near El Barrial (Chihuahua) and along the coast of the Gulf of California in Sonora.

BLACK-TAILED JACKRABBIT (*Lepus californicus*). Abundant in all kinds of open habitats over much of the w. U.S., locally common in the Midwest, sometimes occurs at plague densities in NV and w. UT. Difficult to avoid on night drives. The eastern form is abundant, for example, in Tamarack Ranch SWA (CO), Comanche NG (CO), Sevilleta NWR (NM), Black Kettle NG (OK/TX), and Aubrey Valley (AZ). The western form can be easily seen in Sacramento NWR, Carrizo Plain NMT, Joshua Tree NP (all in CA), and almost everywhere in NV. Forest-adapted populations inhabit the coastal ranges from OR southward; these jackrabbits are particularly abundant at lower elevations of Trinity Alps (CA), where they are easy to see on backcountry roads at night. Animals from low-elevation areas of the sw. U.S. have the biggest ears; jackrabbits living in the area around Yuma (AZ) and in Salton Sea NWR (CA) look like they should be able to fly.

Black-tailed Jackrabbit is a common sight during night drives in the West; it will usually do its best to help you improve your driving skills.

WHITE-SIDED JACKRABBIT (*Lepus callotis*). Very rare and local in the sw. U.S., with only a few sight records from se. AZ and small populations in sw. NM, where it can be seen in open deserts of Animas Valley. Look between miles 34 and 40 on Hwy. 338 south of Animas. Much easier to find in Mexico, for example, around Janos (Chihuahua).

ARCTIC HARE (*Lepus arcticus*). This handsome, often surprisingly tame hare is locally common in the tundra and on rocky slopes in Arctic Canada and Greenland. One reliable place to see it is the rocky peninsula in the town of Churchill (MB). Reportedly common around Pond Inlet (NU) and abundant in Northeast Greenland NP. Rare in Newfoundland, where it can be found on Gros Morne Mt. in Gros Morne NP (NL). Hares living in the far north of the range, for example, in Ellesmere Island NP (NU), remain white year-round. Arctic Hare might be conspecific with **MOUNTAIN HARE** (*L. timidus*), a Eurasian species. Hares living in the tundra of w. AK, previously known as **ALASKA HARE** (*L. othus*), appear to represent an intermediate form. Look for these hares along Teller and Taylor Hwys. near Nome (AK), or anywhere on the Alaska Peninsula.

Snowshoe Hare is the smallest hare in North America, and can sometimes be mistaken for a rabbit.

SNOWSHOE HARE (*Lepus americanus*). Widespread in coniferous forests in the North and in high mountains south to NM; rare in the Appalachians. Numbers in AK and much of Canada fluctuate widely, with peaks about every ten years; the last such peak was in 2009 in AK and YT. If you happen to be in the North when hares are rare, you should still be able to find them by conducting long night drives or by snow tracking in particularly good areas, such as Wood Buffalo NP (NT/AB), Prince Albert NP (SK), the Churchill road system (MB), Riding Mt. NP (MB), Route du Nord (QC), Chic-Choc Mts. (QC), the Trans-Labrador Hwy. (QC/NL), Gros Morne NP (NL), the Yakutat area (AK), the eastern part of Council Hwy. near Nome (AK), or the southern part

At dawn, some hares lay a complex track pattern to prevent predators and hunters from tracking them to their daytime lair.

of Dalton Hwy. (AK). Farther south, look for them at Mt. Evans (CO), at Spruce Knob (WV), in Wakopa WMA (ND), at Sandia Crest (NM), at Spanish Peaks (CO), near Guanella Pass (CO), in Isle Royale NP (MI), in Brule River SF (WI), in Steam Mill Brook WMA (VT), and on Mt. Washington (NH). Hares from some coastal populations in the Pacific Northwest do not turn white in winter; such hares can be seen, for example, in Olympic NP (WA). A blackish form occurs in the Adirondack Mts. (NY). Another dark race inhabits the Sierra Nevada, but is rare and difficult to see; better in years when the snow cover doesn't form until December and already-white hares are easy to spot. Look for it in dense groves of young conifers and streamside thickets around Lake Tahoe (CA/NV).

SCIURIDS (SQUIRRELS, CHIPMUNKS, PRAIRIE DOGS, AND MARMOTS)

Rodents of the squirrel family are unusually diverse in North America, particularly in the West. Most of them are diurnal and very conspicuous, and all are cute, so sciurid watching is a lot of fun. In CA you can see up to ten species in one day by driving across the Sierra Nevada from Mono Lake to the western foothills and back to the sequoia groves. But finding some sciurids requires good planning, as they have small ranges and are active for only two or three months a year.

Chipmunks respond well to squeaking sounds, particularly to "mouse squeakers" sold online and in hunting gear stores. When watching chipmunks in the West, pay attention to their alarm calls and the way they hold their tails while running, because these characteristics can sometimes help tell apart otherwise similar species.

Many ground-dwelling sciurids can live only in places with short grass, where they can see approaching predators. Prairie dogs and marmots solve this problem by trimming the grass themselves; small ground squirrels are sufficiently well camouflaged to hide in the grass; but larger ground squirrels need big mammals to trim the grass for them. Since large wild grazers of the prairie such as American Bison and Columbian Mammoth are now rare or extinct, many ground squirrels use cattle pastures, and are easier to find on heavily grazed lands than in "natural" grasslands. Antelope squirrels often perch on the tops of cacti or desert shrubs during the hottest hours of the day. Flying squirrels are nocturnal, and finding them takes time. Diurnal species are often attracted to bird feeders.

Prairie dog family.

Many sciurids can carry plague, so they should be handled with care. Prairie dogs have been particularly heavily affected since plague was accidentally introduced to North America, and they don't form gigantic towns (colonies) as they did in the past.

NORTHERN FLYING SQUIRREL (*Glaucomys sabrinus*). Very widespread but difficult to see in coniferous forests. Common in AK and Canada, but generally rare in the Lower 48 states, especially in the Appalachians (the Appalachian subspecies is listed as endangered in the U.S.). In Canada it is often seen gliding across forest roads during night drives; in the headlights it looks like a fast-flying white paper envelope. Occurs in sequoia groves in the Sierra Nevada (CA), for example, in Calaveras Big Trees SP and in Tuolumne Grove in Yosemite NP. Other good places to look for it are Groves Prairie Botanical Area in Hoopa Valley (CA), Alfred A. Loeb SP (OR), Mt. Rainier NP (WA), Voyageurs NP (MN), Baxter SP (ME), Algonquin PP (ON), Ashuapmushuan WR (QC), and particularly forests along the northern edge of its range that have 24-hour daylight in early summer, such as the Dawson area (YT) and White Mts. NRA (AK). In the Appalachians, try Grandfather, Roan, and Mitchell Mts. (NC), and Canaan Valley NWR (WV). Very dark-colored squirrels live in Olympic NP (WA) and on Vancouver I. (BC). These and other flying squirrels from the Pacific Coast and the Sierra Nevada might represent a separate species.

SOUTHERN FLYING SQUIRREL (*Glaucomys volans*). Widespread in hardwood forests of the e. U.S. and s. ON. Much easier to see than Northern Flying Squirrel. Look for it, for example, around Cades Cove in Great Smoky Mts. NP, in sw. Cape Cod (MA), in Shenandoah NP (VA), in Muscatatuck NWR (IN), in Pee Dee NWR (NC), or at Archbold Biological Station (FL), where it often nests in sheds and garages. In winter, it comes to feeders at Long Branch Nature Center in Arlington (VA) about an hour after sunset; call the center for viewing information. In Algonquin PP (ON), where both flying squirrels are common, this species can be distinguished from Northern Flying Squirrel in the fall months by gliding more horizontally, since it puts on less fat.

Southern Flying Squirrel is easy to see in winter in Long Branch Nature Center in Arlington, Virginia.

PINE SQUIRREL (*Tamiasciurus hudsonicus*). Formerly known as **RED SQUIRREL**, this lively rodent is abundant in fir, spruce, hemlock, and redwood forests over much of N. America, and increasingly common in other forest types, particularly in the ne. U.S. Easy to see, for example, at the higher elevations of Great Smoky Mts. NP (TN/NC), as well as in Voyageurs NP (MN), on Isle Royale (MI), in Yellowstone NP (WY/MT/ID), and in almost any forest in n. New England, Canada (including Vancouver I.), and AK. The distinctive southwestern form, sometimes called **SPRUCE SQUIRREL**, is common in the mountains of CO, NM, and UT; look for it in the Sandia Mts. (NM), Uinta Mts. (UT), and Rocky Mt. NP (CO). Rare on the mountaintops of s. AZ; the subspecies occurring on Mt. Graham (AZ) is listed as endangered in the U.S. The red-bellied form previously known as **DOUGLAS'S SQUIRREL** (*T. douglasii*) is very common on the mainland Pacific Coast, including Garibaldi PP (BC), Olympic NP (WA), Redwood NP (CA), and in the Sierra Nevada.

ABERT'S SQUIRREL (*Sciurus aberti*). Also known as **TASSEL-EARED SQUIRREL**, it is uncommon and local in ponderosa pine forests of the sw. U.S. The gray form can be seen, for example, in Walnut Canyon NMT, along Hwy. 191 south from Eagar, at Mt. Lemmon, at the South Rim of the Grand Canyon, along the nature trail at Little America Hotel in Flagstaff, and at Los Burros Campground in Apache-Sitgreaves NFs (all in AZ); near Devils Canyon Campground in Manti-La Sal NF (UT); in Bandelier NMT, Sugarite Canyon SP, Water Canyon, and at Sandia Crest (all in NM); and along Animas Mt. Trail near Durango (CO). It is also common at Mt. Graham (AZ), where it has been introduced. The black form is common in the mountains above Boulder (CO), where brown individuals also occur, in Elk Meadow Park (CO), and in Cimarron Canyon SP (NM). The rare, beautiful white-tailed form (**KAIBAB SQUIRREL**) occurs on Kaibab Plateau (AZ), where it can be seen (usually with considerable effort) around Jacob Lake and DeMotte Campgrounds and along Mt. Trumbull Trail.

MEXICAN FOX SQUIRREL (*Sciurus nayaritensis*). Locally common in the Chiricahua Mts. (AZ), especially along the bottoms of forested canyons. Look for it in Cave Creek Canyon and Chiricahua NMT. In some years this squirrel becomes difficult to find, probably because of yet undiscovered local migrations.

EASTERN FOX SQUIRREL (*Sciurus niger*). Widespread but local in the East and the Midwest; occurs along riparian corridors west to CO and MT. Introduced in many cities in the West, and now abundant, for example, in Davis (CA) and Boulder (CO). Prefers open forests with oaks, hickories, and pines. Numerous distinctive color variants occur in different parts of the range. Northern races tend to be smaller and less brightly colored. The gray-and-rusty form is common, among other

Both Eastern Fox and Eastern Gray Squirrels can be black in some areas.

places, in Pictured Rocks NLS (MI), Muscatatuck NWR (IN), Chinco-teague NWR (VA), Wichita Mts. WR (OK), and Heyburn WMA (OK). The silver-colored race (**DELMARVA FOX SQUIRREL**) is common in St. Mary's River SP and Blackwater NWR (both in MD); this race is listed as endangered in the U.S. The gray black-headed form occurs in Raven Rock SP (NC) and Congaree NP (SC). The rusty-and-orange southern form can be seen in Kisatchie NF (LA); bright orange animals occur in Barataria PR (LA); orange white-nosed ones live in Di-Lane Plantation WMA (GA). The southwestern form is easy to see, for example, at Belton Lake and in Big Thicket NPR (both in TX). The

GRAY-COLLARED CHIPMUNK (*Tamias cinereicollis*). Occurs in summer and fall in mountain forests of sw. NM and e.-cen. AZ, particularly at or above the pine-fir ecotone. Good places to look for it include the Arboretum at Flagstaff, the nature trail at Little America Hotel in Flagstaff, Walnut Canyon NMT, Hwy. 191 south from Eagar, Chiricahua NMT, Woods Canyon Lake, Los Burros Campground in Apache-Sitgreaves NFs (all in AZ), Sacramento Mts. (NM), and the highest peaks of Magdalena Mts. (NM).

CLIFF CHIPMUNK (*Tamias dorsalis*). Locally common in wooded rocky areas of the arid sw. U.S., particularly in canyons with piñon and juniper. Easy to see in City of Rocks NR (ID), in Zion and Capitol Reef NPs (UT), at both rims of the Grand Canyon (AZ), in Chiricahua NMT (AZ), in Hickison Petroglyph RA (NV), at Mt. Graham (AZ), in Cave Creek Canyon Ranch (AZ), in El Malpais NMT (NM), and in Water Canyon (NM).

HOPI CHIPMUNK (*Tamias rufus*). Common on rocky slopes with piñon-juniper woodland throughout the Colorado Plateau, mostly at low elevations. Easy to find from early spring to late fall in Arches and Canyonlands NPs, in Natural Bridges NMT, at the bases of cliffs in Valley of the Gods (all in UT), and in Colorado NMT (CO).

COLORADO CHIPMUNK (*Tamias quadrivittatus*). Very common from spring to fall in pine and juniper forests in the mountains of CO, and at high elevations in n. NM and se. UT. Good places to see it include Natural Bridges NMT (UT), viewpoint pullouts in Rocky Mt. NP (CO), bristlecone pine groves at Mt. Evans (CO), Wheeler Peak (NM), Sandia Crest (NM), and Aguirre Springs Campground near Las Cruces (NM).

UINTA CHIPMUNK (*Tamias umbrinus*). Locally common in the mountains of the interior w. U.S., mostly in open coniferous forests with lots of fallen timber above 7,000 ft. (2,100 m). Look for it from spring to late fall in Great Basin NP (NV), at Bear Creek Summit (NV), in Cherry Springs NA (ID), in Bryce Canyon NP (UT), in Uinta Mts. (UT), along Mt. Washburn Trail in Yellowstone NP (WY), and in the hills above Boulder (CO). Uncommon and local in the s. Sierra Nevada, where it occurs in open forests just below the timberline, at 7,500 to 10,500 ft. (2,200–3,150 m); try the eastern slopes of Mt. Whitney (CA).

PALMER'S CHIPMUNK (*Tamias palmeri*). Endemic to Spring Mts. (NV), where it's most common around rocks and large logs at about 8,000 to 12,000 ft. (2,400–3,600 m). Tame chipmunks are usually present in summer and early fall around Las Vegas Ski and Snowboard Resort, and around Mt. Charleston Lodge. Possibly a subspecies of Uinta Chipmunk.

PANAMINT CHIPMUNK (*Tamias panamintinus*). Locally common along the CA/NV state line on rocky slopes with piñon and juniper at 5,000 to 9,000 ft. (1,500–2,700 m). Occurs along the upper portion of Mono Pass Trail in Yosemite NP, in South Sierra Wilderness (CA), Ancient Bristlecone Pine Forest in Inyo NF (CA), along the trail to Telescope Peak in Death Valley NP (CA), near Montgomery Pass (NV), and at Deer Creek Picnic Area on Hwy. 158 in Spring Mts. (NV).

LEAST CHIPMUNK (*Tamias minimus*). Widespread in much of interior Canada and the arid w. U.S., but uncommon in its Canadian range. Inhabits a broad variety of habitats, from alpine meadows and montane forests to deserts. This little gem is highly variable in coloration; even animals from adjacent populations can look very different. Common from early spring to late fall, for example, in Lassen Volcanic NP (CA), in Yellowstone and Grand Teton NPs (WY), in Great Basin NP (NV), in City of Rocks NR (ID), in Cherry Springs NA (ID), along Owyhee Uplands Back Country Bwy. (ID/OR), in Dinosaur NMT (CO/UT), around the headquarters of Arapaho NWR (CO), around Taos (NM), in Cimarron Canyon SP (NM), at Wildhorse Crossing Campground (NV), in Rockport SP (UT), in Bryce Canyon NP (UT), and in Gilbert-Baker WMA (NE). In the Sierra Nevada it occurs mostly on sagebrush-covered eastern slopes at 7,000 to 10,000 ft. (2,100–3,000 m); look for it near Taylor Creek Visitor Center at Lake Tahoe, in the hills above Mono Lake, and along the eastern approach to Tioga Pass (all in CA). In summer it's common above the timberline at Mt. Evans (CO). Particularly light-colored chipmunks live in Badlands NP (SD). The distinctive northwestern form can be seen in summer at mile 61 (km 99) of Icefields Pkwy. in Jasper NP (AB). The bright-colored northeastern race occurs in Voyageurs NP (MN) and Pukaskwa PP (ON); it seems to be declining in many areas, possibly because it's being replaced by Eastern Chipmunk, which is expanding its range as the climate warms up. The black-striped subspecies from s.-cen. NM has become very rare, and might be listed as endangered in the U.S. by the time you read this book; it now survives only on the n. slope of Sierra Blanca Peak near Sunspot.

ALPINE CHIPMUNK (*Tamias alpinus*). A high-elevation endemic of the Sierra Nevada. Although tiny, it is really hardy; look for it in June–October on rocky and talus slopes from 8,000 to 12,000 ft. (2,400–3,600 m), north to Saddlebag Lake near Tioga Pass. Good places include a talus slope three mi. up Mono Pass Trail in Yosemite NP, Mount Whitney area, and Mineral King in Sequoia NP (all in CA).

LONG-EARED CHIPMUNK (*Tamias quadrimaculatus*). Common in the cen. Sierra Nevada from 4,000 to 7,000 ft. (1,200–2,100 m), particularly in small clearings in mature coniferous forests. Look for it from

Alpine Chipmunk is the smallest member of the squirrel family in North America, and one of the most beautiful and friendly. It is endemic to the highest portions of the Sierra Nevada. An intrepid hiker will be rewarded by a chance to observe this little gem up close in its inhospitable habitat.

late spring to fall around Lake Tahoe (CA/NV), in Grover Hot Springs SP (CA), and in Tuolumne Grove in Yosemite NP (CA).

LODGEPOLE CHIPMUNK (*Tamias speciosus*). Common in the Sierra Nevada, mostly from 6,000 to 9,000 ft. (1,800–2,700 m) in open coniferous forests and sequoia groves. Look for it from early spring to late fall around Lake Tahoe (CA/NV), as well as in Yosemite, Kings Canyon, and Sequoia NPs (all in CA). Local in the mountains of s. CA above 7,000 ft. (2,100 m); try looking for it along Angeles Crest Hwy. (Hwy. 2) in Mt. San Jacinto SP, and in San Gabriel Mts. NMT.

CALIFORNIA CHIPMUNK (*Tamias obscurus*). Locally common in dry piñon-juniper forests of San Bernardino and San Jacinto Mts. (CA), especially around rocky outcrops and talus slopes. Most active in late summer and early fall. Look for it in Pine City area of Joshua Tree NP, Cuyamaca Rancho SP, San Jacinto SP, and Santa Rosa and San Jacinto NMT (all in CA).

MERRIAM'S CHIPMUNK (*Tamias merriami*). Common in s. CA in chaparral and open woodland on rocky slopes, up to 4,500 ft. (1,350 m). Reliably found along Cone Peak Trail in Santa Lucia Mts., on Loma Prieta Mt. in Santa Cruz Mts., in juniper-covered hills at the periphery of Carrizo Plain NMT, in shady wooded areas in Pinnacles NP, and at lower elevations in Sequoia and Kings Canyon NPs. The distinctive southern form can be seen around Chilao Visitor Center in San Gabriel Mts. NMT, and at higher elevations in Cuyamaco Ranco SP (all in CA).

SONOMA CHIPMUNK (*Tamias sonomae*). Locally common in nw. CA in chaparral and brushy clearings in coniferous forests below 5,000 ft. (1,500 m). It is usually absent from redwood forests and the coast; however, it is common in Muir Woods NMT and present in Point Reyes NSS. Also common in Shasta-Trinity NF and Mendocino NF (all in CA).

YELLOW-CHEEKED CHIPMUNK (*Tamias ochrogenys*). Common in redwood forests of coastal n. CA, from Humboldt Redwoods SP (where abundant) to Armstrong Redwoods SR. A good place to see it is at the bird feeders behind the dining hall at the Biological Field Station in Albion (CA).

ALLEN'S CHIPMUNK (*Tamias senex*). Locally common from late spring to late fall in chaparral and moist coniferous forests in n. CA and s.-cen. OR; rare in the Sierra Nevada south to Yosemite NP (CA) from 5,000 to 9,000 ft. (1,500–2,700 m). Occurs in dense, shady forest along Echo Lake Rd. near Lake Tahoe (CA/NV), in Trinity Alps (CA), and in Cascade-Siskiyou NMT (OR).

SISKIYOU CHIPMUNK (*Tamias siskiyou*). Occurs between Klamath and Rogue Rs. in nw. CA and s. OR, mostly in coniferous forests from 3,300 to 6,600 ft. (1,000–2,000 m). Look for it from spring to late fall in Jedediah Smith Redwoods SP (CA), in Smith River NRA (CA), in Kalmiopsis Wilderness (OR), and in Oregon Caves NMT (OR) near the visitor center.

TOWNSEND'S CHIPMUNK (*Tamias townsendii*). Uncommon in coastal forests from the lower Fraser R. (BC) to OR, mostly in dense forests with lots of undergrowth. Good places to find it include Olympic NP (WA) and Devils Elbow SP (OR). The grayish inland form occurs in Mt. St. Helens NMT (WA) and at Mt. Hood (OR), but is usually difficult to find.

RED-TAILED CHIPMUNK (*Tamias ruficaudus*). Local and uncommon in parts of the n. Rockies, mostly in clearings and dense shrubbery in coniferous forests. Look for it in summer and early fall at Avalanche Lake in Glacier NP (MT), at Metaline Falls (WA), in Kootenai NWR (ID), and around Lake Coeur d'Alene (ID). Occurs in Waterton Lakes NP (AB).

Townsend's Chipmunk inhabits temperate rainforests of the Pacific Northwest.

HARRIS'S ANTELOPE SQUIRREL (*Ammospermophilus harrisii*). Common in sparsely vegetated lowland deserts in s. AZ; local in NM. Often the only mammal visible in the heat of the day in Organ Pipe Cactus NMT, Saguaro NP, San Bernardino NWR, along the road to Cave Creek Canyon (all in AZ), and around Red Rock (NM).

TEXAS ANTELOPE SQUIRREL (*Ammospermophilus interpres*). Uncommon in foothill deserts of Rio Grande Valley (NM) and far w. TX. Tends to be less approachable than other antelope squirrels. Look for it in rocky desert washes in Sevilleta NWR, around Placitas, around Aguirre Springs Campground in Organ Mts.-Desert Peaks NMT (all in NM), and in Big Bend NP (TX).

WHITE-TAILED ANTELOPE SQUIRREL (*Ammospermophilus leucurus*). Abundant in creosote bush–dominated deserts of the sw. U.S., except for s. AZ and most of NM. Easy to find, for example, in Joshua Tree NP, Anza-Borrego Desert SP, Mojave NPR, Providence Mts. (all in CA), Fish Springs NWR (UT), Chaco Culture NHP (NM), Grand Staircase-Escalante NMT (UT), Arches NP (UT), and Valley of Fire SP (NV).

NELSON'S ANTELOPE SQUIRREL (*Ammospermophilus nelsoni*). Occurs almost exclusively within Carrizo Plain NMT (CA), where it's relatively common in arid grasslands on the valley floor; the best way to find it is to scan the brushy patches in the grasslands with a birding

Nelson's Antelope Squirrel is an endangered species: most of its habitat has been destroyed by agricultural development in the Central Valley of California.

scope on a spring morning. It used to be particularly numerous around the visitor center, but has become rare there by 2014, probably because of the ongoing drought; try the southern end of the valley. Also present but rare and difficult to find in Tule Elk SR and along Little Panoche Rd. (both in CA).

GOLDEN-MANTLED GROUND SQUIRREL (*Callospermophilus lateralis*). Often mistaken for a chipmunk, this colorful rodent is very common in summer and early fall in the mountains of the West, from BC to NM. Prefers rocky habitats, particularly at or above the timberline. Signs that say "Do not feed wildlife" at roadside viewpoints and campsites in many national parks of the West refer mostly to this friendly animal. Easy to see in Lassen Volcanic NP (CA), in all NPs of the Sierra Nevada (CA), in Ancient Bristlecone Pine Forest (CA), anywhere in the Cascade Mts. of OR, at Quemado Lake (NM), at Redfish Lake (ID), in Yellowstone NP (WY/ID/MT), on both rims of the Grand Canyon (AZ), in Grand Staircase-Escalante NMT (UT), and almost everywhere in the Rockies from Banff NP (AB) to Wheeler Peak (NM).

CASCADE GROUND SQUIRREL (*Callospermophilus saturatus*). Common in the Cascades of s. BC and WA, mostly at or above the timberline. Easy to see in early summer in Skagit Valley PP (BC); North Cascades NP (WA); Mt. Rainier NP (WA), especially in the picnic area behind Sunrise Lodge and at Frozen Lake lookout; also in Mt. St. Helens NMT (WA) and Conboy Lake NWR (OR).

MERRIAM'S GROUND SQUIRREL (*Urocitellus canus*). Also known as **COLUMBIA PLATEAU GROUND SQUIRREL**, it is active from late spring to midsummer in sagebrush flats from ne. CA to w. ID. Common in e. OR. Good places to see it include Butte Valley NG (CA), Hart Mt. NWR (OR), Owyhee Uplands Back Country Bwy. (ID/OR), Deer Flat NWR (ID), Sheldon NWR (NV), and Jack's Creek 20 miles south of Boise (ID). The distinctive eastern subspecies occurs in Hagerman Fossil Beds NMT (ID).

COLUMBIAN GROUND SQUIRREL (*Urocitellus columbianus*). Locally common in grasslands, meadows, and shrublands in and around the n. Rockies. Easy to find from late spring to midsummer in Banff NP (AB), in Manning PP (BC), along Hidden Lake Overlook Trail and at Avalanche Lake in Glacier NP (MT), in Lee Metcalf NWR (MT), in Payette NF (ID), in Silver Creek Preserve (ID), around Fairfield (ID), around Enterprise (OR), along Red Sleep Mt. Dr. in National Bison Range (MT), and at the rest stop on westbound I-90 a few miles east of the MT/ID state line.

WYOMING GROUND SQUIRREL (*Urocitellus elegans*). Fragmented populations of this uncommon species occur in grasslands, meadows, and sagebrush deserts from n. NV to sw. KS. It often forms mixed colonies with prairie dogs. Occurs at lower elevations in Rocky Mt. NP (CO), in Owyhee Mts. (ID), at Steens Mt. (OR), in Red Rock Lakes NWR (MT), around the headquarters of Arapaho NWR (CO), at Kenosha Pass (CO), in Florissant Fossil Beds NMT (CO), in Fossil Butte NMT (WY), in Hutton Lake NWR (WY), at Como Bluff Dinosaur Graveyard (WY), around Dillon (MT), and along the Cimarron R. in KS.

RICHARDSON'S GROUND SQUIRREL (*Urocitellus richardsonii*). Locally common in late spring and early summer in the grasslands of the northern part of the prairie zone, particularly in overgrazed pastures where it is very easy to find. Conspicuous colonies of this species can be seen in City of Rocks NR (ID), Bowdoin NWR (MT), Missouri Headwaters SP (MT), Glacier NP (MT), Medicine Lake NWR (MT), and Elk Island NP (AB). Semi-tame squirrels live on the grounds of Zoo Montana in Billings (MT). The increasingly rare sw. subspecies can still be found in Sheldon NWR (NV).

WASHINGTON GROUND SQUIRREL (*Urocitellus washingtoni*). Rare and local in sandy grasslands and sagebrush deserts of the Columbia Plateau. Small colonies are active from early spring to early summer in Channeled Scablands, in the Seep Lakes Unit of Columbia Basin WMA, around Wenatchee, in public fishing areas off Rte. 262 near O'Sullivan Dam, in Columbia NWR (all in WA), and in Umatilla NWR (OR). This squirrel tends to be absent from many seemingly appropriate habitats, and finding it might take a lot of effort, but spring grasslands are not the worst place to be driving around.

THIRTEEN-LINED GROUND SQUIRREL (*Ictidomys tridecimlineatus*). This strikingly beautiful little squirrel is abundant from spring to fall throughout the prairie zone and east to OH in areas with short grass. Despite being so common and relatively tame, it is annoyingly difficult to see, although some individuals become habituated to people and handouts. Look for it (best in the morning) at freeway rest areas anywhere within its range, near the park office at Voyageurs NP (MN), on the grounds of the Denver Zoo (CO), at the edges of prairie dog towns around Boulder (CO), in Grasslands NP (SK), Thunder Basin NG (WY), Theodore Roosevelt NP (ND), Cimarron NG (KS), Attwater Prairie Chicken NWR (TX), High Cliff SP (WI), Pipestone NMT (MN), Kalsow Prairie SPR (IA), Tippecanoe River SP (IN), DeSoto NWR (IA), and in the vicinity of Oxford (OH).

MEXICAN GROUND SQUIRREL (*Ictidomys mexicanus*). Uncommon and local in se. NM and w. TX. Occasionally seen along highways, where it feeds on roadkill. Occurs in Balmorhea SP, Seminole Canyon SHP, Bentsen-Rio Grande Valley SP, San Angelo area, and Laguna Atascosa NWR (all in TX). Tame individuals are easy to see on the New Mexico State University campus in Las Cruces (NM) and in the town of Carlsbad (NM).

MOHAVE GROUND SQUIRREL (*Xerospermophilus mohavensis*). Very rare and local in the w. Mojave Desert. Active from mid-spring to midsummer in sandy areas with lots of creosote bush, for example, in Desert Tortoise NA (CA). Note that ground squirrels seen along roads in this area are usually California Ground Squirrels; seeing the elusive Mohave Ground Squirrel takes a lot of walking.

SPOTTED GROUND SQUIRREL (*Xerospermophilus spilosoma*). Uncommon and generally difficult to see in sandy areas from w. NE to AZ and s. TX; usually active only in morning hours. Look for it at Grand Canyon NP Airport (AZ), in Big Bend Ranch SP (TX), at Aguirre Springs Campground (NM), on New Mexico State University campus in Las Cruces (NM), in Como Bluff Dinosaur Graveyard (WY), and in Canyons of the Ancients NMT (CO). Very common in parts of Mexico, for example, at Teotihuacan Ruins (Mexico DF) from June till December.

ROUND-TAILED GROUND SQUIRREL (*Xerospermophilus tereticaudus*). Locally common from late winter to midsummer in sandy deserts with lots of creosote bush in parts of the sw. U.S. Occurs in Organ Pipe Cactus NMT; in Saguaro NP; in Lost Dutchman SP; on the grounds of the Phoenix Zoo, Papago Park, and Desert Botanical Gar-

OLYMPIC MARMOT (*Marmota olympus*). Endemic to high-elevation meadows and talus slopes in Olympic NP (WA). Although the population is very small, this marmot can be reliably seen in summer and early fall at Hurricane Ridge around the visitor center.

VANCOUVER ISLAND MARMOT (*Marmota vancouverensis*). Inhabits steep slopes with subalpine meadows on s. Vancouver I. (BC). This beautiful marmot is very rare, but a few are usually present in summer on Mt. Modeste and around Mt. Washington Alpine Ski Resort (try Hawk Chairlift around pylon 13). Listed as endangered in Canada.

HOARY MARMOT (*Marmota caligata*). Locally common from AK, where it occurs down to the sea level, to WA, ID, and MT, where it lives above the timberline on alpine meadows and talus slopes. Occasionally lives in towns, for example, in Gustavus (AK), where black individuals can sometimes be seen. Common in summer in Denali NP (AK) and along Harding Icefield Trail in Kenai Fjords NP (AK). A darker race inhabits the northeastern part of the range; look for these animals at Teslin Lake (YT), in the Ogilvie Mts. (YT), on Pink Mt. (BC), and on Mt. Robson (BC). Pale-colored subspecies can be easily seen (also in summer) in North Cascades NP (WA), on Mt. Rainier (WA), and in Glacier NP (MT).

Like most marmots, Hoary Marmot is highly variable in color.

Woodchuck is the only marmot of the East.

ALASKA MARMOT (*Marmota broweri*). Endemic to Brooks Range in AK, where it lives on talus slopes near meadows. The easiest place to see it is at mile marker 259 on Dalton Hwy., where it can be found within a short hike from the road (climb to the fork of the Y-shaped valley on the steep mountain slope to the east). Similar-looking marmots occur on mountain passes along Dempster Hwy. (YT), but it is unclear if they belong to this species or the previous one.

WOODCHUCK (*Marmota monax*). Also known as **GROUNDHOG** and **WOOD MARMOT**, it is common but seldom seen in forests and clearings of interior AK, most of Canada, and much of the e. U.S. In some areas, particularly in VA, WV, OH, PA, and ON, it lives also in towns and along highways, and is much easier to see there. For nonferal woodchucks, look from early spring to late fall in Stone Mt. PP (BC), in Riding Mt. NP (MB), in Algonquin PP (ON), in Shenandoah NP (VA), in lowland areas of Great Smoky Mts. NP (TN/NC), in Erie NWR (PA), in Lime Rock Preserve (RI), and in Rock Cut SP (IL). Animals from the interior U.S. tend to be more reddish, while those in interior Canada are darker. The largest individuals, with beautiful silver-black fur, occur in the vicinity of Goose Bay (NL). The most colorful Woodchucks, cinnamon-colored, live in the northwestern part of the range; this race is generally rare, but can be found with some luck along Klondike Hwy. (YT) and around Eagle (AK).

POCKET GOPHERS

Pocket gophers are an ancient family, unique to North and Central America. They are usually easy to find (their mounds are very conspicuous), but difficult to see as they spend most of the time underground. Of the three genera occurring in North America, *Thomomys* species tend to show up on the surface more often, especially at night or on overcast days. Once you have found a good colony, the waiting time needed to see a *Thomomys* is usually from a few minutes to a few hours. Gophers in some colonies are consistently easier to see than in others. *Geomys* and *Cratogeomys* species are more strictly subterranean, and normally leave their burrows only at night. A good way to see a *Geomys* is to park your car overnight facing a group of active burrows, and briefly turn the lights on every 10 to 15 minutes. If the snow cover exceeds 8 to 10 in. (20–25 cm), pocket gophers usually switch to tunneling through the snow and cease to appear on the surface. Pocket gophers can be live-trapped, but it takes some effort.

Dispersing juveniles travel on the surface and can get into all kinds of trouble. They are sometimes seen crossing roads, usually at night, and can enter rodent traps set far from gopher burrows. Some accidentally get caught in concrete drains, gas pipe manholes, and under cattle guards.

Many species are highly variable in appearance, with consistent differences even between adjacent colonies. Fans of the phylogenetic species concept (PSC) should ignore all other animals and concentrate on watching pocket gophers, as there are hundreds of undescribed PSC species of these rodents in the U.S.

BOTTA'S POCKET GOPHER (*Thomomys bottae*). The most common pocket gopher in much of the sw. U.S. Abundant, for example, around Lucia, in Carrizo Plain NMT, in the town of Elk, at lower elevations in Joshua Tree NP, and in virtually all protected areas of the Central Valley (all in CA); also on the floor of Valles Caldera (NM), and in the flower beds near the visitor center in Zion NP (UT). One place where gophers are particularly easy to see on the surface is Cesar Chavez Park in Berkeley (CA). A reddish-colored form occurs in Big Bend NP (TX), while very dark animals inhabit the coast of s. OR. There are many other distinctive races, such as the one occurring in Fish Springs NWR (UT). The large northern form formerly known as **TOWNSEND'S POCKET GOPHER** (*T. townsendi*) occurs at isolated locations from ne. CA to s. ID, mostly in arid river valleys. Look for it around Eagle Lake (CA), near Winnemucca (NV), and in Bruneau Dunes SP (ID). The small southern subspecies formerly known as **ANIMAS MOUNTAINS** or **SOUTHERN POCKET GOPHER** (*T. umbrinus*) inhabits desert grasslands and pine-oak forests in s. AZ and NM; it is common, for example, in Buenos Aires NWR and in Carr and Madera Canyons (all in AZ).

CAMAS POCKET GOPHER (*Thomomys bulbivorus*). Locally common in well-drained areas of Willamette Valley (OR), mostly on agricultural lands. Easy to see in and around Ankeny NWR, particularly in sum-

Botta's Pocket Gopher. Gophers living in sparsely vegetated areas have to show up on the surface more often.

mer. Being mostly nocturnal, it doesn't tolerate white light; you'll find it easier to observe this gopher if you cover your light source with red film.

WESTERN POCKET GOPHER (*Thomomys mazama*). Alternatively known as **MAZAMA POCKET GOPHER**, this species is uncommon and local in the nw. U.S., where it occurs in many types of habitats. It seems to spend more time on the surface than other pocket gophers. Look for it along Pit River (CA) and in Oregon Dunes NRA (OR). Isolated populations in WA have been described as separate subspecies; they are severely fragmented, and some have gone extinct. You can see these races along the road to Hoh Rain Forest in Olympic NP (WA), and on the airfields of some regional airports, such as Olympia Airport (if you are planning to watch the airfield through the fence at night, it's a good idea to notify the airport security in advance to avoid misunderstandings). Mima Mounds NA PR (WA) protects a landscape of huge mounds created over millennia by many generations of Western Pocket Gophers.

MOUNTAIN POCKET GOPHER (*Thomomys monticola*). Occurs mostly in the Sierra Nevada, particularly at the edges of subalpine meadows and in overgrown forest clearings. Common along Tioga Rd. in Yosemite NP (CA), in lawns and meadows around Lake Tahoe (CA/NV), and on the lower slopes of Mt. Shasta (CA). In areas frequented by

Dispersing juvenile Mountain Pocket Gopher.

tourists, these gophers sometimes get used to people and become more approachable in late summer and early fall.

NORTHERN POCKET GOPHER (*Thomomys talpoides*). Common in colder parts of the w. U.S. and locally in sw. Canada, mostly in open areas with meadows and streams, but also in fields, grasslands, and forests. Look for it, for example, at higher elevations in Valles Caldera NPR (NM), in City of Rocks NR (ID), in Yellowstone NP (MT/ID/WY), in Rocky Mt. NP (CO), in Hanford Reach NMT (WA), in Grasslands NP (SK), in Elk Island NP (AB), and in all NPs in SD and ND, particularly in Badlands NP (SD). On the summit plateau of Grand Mesa (CO), these gophers spend a lot of time aboveground in early spring; while in Yellowstone NP they are often active on the surface on warm October evenings (try the slope below Lake Hotel).

IDAHO POCKET GOPHER (*Thomomys idahoensis*). This small, blonde, rather pretty rodent is probably the most difficult N. American pocket gopher to see. It is uncommon and local in gently sloping meadows with sparse grass, saltbrush, and hard soil from sw. MT to sw. WY. Red Rock Lakes NWR (MT), Fossil Butte NMT (WY), the vicinity of Fort Bridger SHS (WY), and the vicinity of Atomic City (ID) are good places to try looking for it.

WYOMING POCKET GOPHER (*Thomomys clusius*). Inhabits one small area of s.-cen. WY, where it prefers dry upland meadows covered with greasewood. Look for it along the unpaved road branching north off

I-80 at exit 173, toward Jeffrey City. Early winter is the best time, because in summer this species is almost strictly nocturnal.

DESERT POCKET GOPHER (*Geomys arenarius*). Very local in sandy lowlands of s.-cen. NM and far w. TX. This species virtually disappears in years following droughts, and was impossible to find in 2012. In 2004 it was common at the periphery of White Sands NMT (NM), at Gran Quivira Ruins in Salinas Pueblo Missions NMT (NM), and along the shores of the Rio Grande near Mesquite (NM).

ATTWATER'S POCKET GOPHER (*Geomys attwateri*). Common in dry, open plains of coastal TX around Corpus Christi; abundant in Aransas NWR. Seldom appears on the surface; plan to spend all night waiting for it to emerge. I'd recommend not doing this in summer when it's too hot to keep the car windows up and there are too many biting insects to keep them down.

BAIRD'S POCKET GOPHER (*Geomys breviceps*). Occurs in areas with sandy soils in much of e. TX and locally in se. OK, sw. AR, and w. LA. Look for it, for example, in Davy Crockett NF, in Anahuac NWR, anywhere around Houston (all in TX), and in pine forests in the westernmost part of Kisatchie NF (LA).

PLAINS POCKET GOPHER (*Geomys bursarius*). Common in sandy areas through much of the prairie zone. Occurs in almost all prairie NGs, for example, in Sheyenne NG (ND), Buffalo Gap NG (SD), Black Kettle NG (OK/TX), McClellan Creek NG (TX), Oglala NG (NE), and Pawnee NG (CO). Other good places to see it include Quivira NWR (KS), Caverns of Sonora area (TX), Big Sandy Creek Trail in Big Thicket NPR (TX), the outskirts of Dexter (NM), the Clinton Lake area (IL), Kalsow Prairie SPR (IA), Agassiz NWR (MN), Little River NWR (OK), and the vicinity of Emerson (MB).

JONES'S POCKET GOPHER (*Geomys knoxjonesi*). Uncommon and local in sandy areas with sparse vegetation in extreme se. NM and adjacent w. TX. Usually relatively easy to find at the lowest elevations of Guadalupe Mts. NP (TX) and on heavily grazed sandy pastures around Jal (NM).

LLANO POCKET GOPHER (*Geomys texensis*). Abundant in a small area of TX west of San Antonio. Look for it in Lost Maples SNA in sandy patches. In 1998, a few gophers lived near the campground there and could be easily seen at dusk. Reportedly also common in the nearby Hill Country SNA.

Molehills (left) usually look like symmetrical cones, while pocket gopher mounds (right) tend to be fan-shaped, with an entrance plug sometimes visible on top. Moles dig only with their feet, but pocket gophers also use their teeth and can tunnel through much harder, drier soil.

TEXAS POCKET GOPHER (*Geomys personatus*). Common in sandy lowlands in parts of s. TX; easy to find in Laguna Atascosa NWR and around Corpus Christi. A dark-colored race occurs around Pinto in Kinney Co. Another distinctive race, occurring around Winter Haven in Dimmit Co., has been proposed to represent a separate species, **STRECKER'S** or **CARRIZO SPRINGS POCKET GOPHER** (*G. streckeri*).

SOUTHEASTERN POCKET GOPHER (*Geomys pinetis*). Locally common in sandy areas of s. AL and GA; abundant in much of n. FL. Gophers in Withlacoochee SF (FL), Wekiwa Springs SP (FL), and Fort Frederica NMT (GA) seem to be relatively easy to see aboveground. In Conecuh NF (AL) some gophers have beautiful golden fur. A distinctive form occurs in Cumberland Island NSS (GA).

YELLOW-FACED POCKET GOPHER (*Cratogeomys castanops*). Locally common in open grasslands from se. CO to s. TX; it apparently prefers clay and rocky soils to sand. Often can be observed from blinds installed at prairie chicken leks. Easy to find in Lake Rita Blanca SP (TX), Cimarron NG (KS), and Comanche NG (CO); occurs in Big Bend NP (TX).

HETEROMYIDS
(POCKET MICE AND KANGAROO RATS)

A family of seed-eating nocturnal dwellers of arid habitats, heteromyids are most diverse in the w. U.S. and in Mexico, although a few species occur in Canada, in the Midwest, and in tropical forests south to Ecuador. They are usually relatively easy to trap (except when natural food is abundant), but almost all species are also easy to see by spotlighting. Kangaroo rats will often continue to browse when lit by a flashlight beam. Very small species can be unbelievably tame, and can sometimes be picked up by hand and even hand-fed. (Interestingly, some very small species of *Mus* mice in African savannas and pygmy jerboas of Central Asia are also unusually tame and easy to handle if found at night.) Being able to observe the natural behavior of these charming creatures in the wild always feels like a privilege.

Heteromyids are usually less active around the full moon and on very cold or hot nights. Smaller species often remain in their burrows all winter. In the deserts of the sw. U.S. they are often the most plentiful mammals, but their numbers can change dramatically from year to year. Most species prefer sandy soils, but some can live in rocky or clay deserts. Some kangaroo rats build large domelike mounds. If you approach an occupied mound very quietly and drum or tap on it, they will sometimes drum back.

All heteromyids can hop, and will take long leaps when escaping danger. This gait helps distinguish pocket mice from other desert rodents when they cross the road in front of your car, but in the prairies you have to tell them from jumping mice, which also hop.

Although heteromyids can be unbelievably abundant in some areas, and can often feed in agricultural fields, many species are adapted to using particular types of soil for burrowing and sand-bathing, so they are extremely sensitive to habitat modification by humans. Numerous species and subspecies, almost all of them in the southern half of CA, are now endangered, threatened, or declining. If you are planning to go looking for them in parks or nature reserves, it is a good idea to consult with the park naturalist first.

MEXICAN SPINY POCKET MOUSE (*Liomys irroratus*). The northernmost representative of a large tropical subfamily, this funny-looking rodent is locally common in dense brush, palm groves, and prickly pear patches in extreme s. TX. Easy to find in Laguna Atascosa NWR, where it's often active at dusk, particularly in winter. Recent data suggest that its scientific name should be *Heteromys irroratus*.

HISPID POCKET MOUSE (*Chaetodipus hispidus*). Abundant in sparsely vegetated areas in the shortgrass prairie zone and in parts of the sw. U.S., particularly on sandy soils. Easy to find, for example, in Guadalupe River SP (TX), in Carlsbad Caverns NP (NM), in Wichita Mts. WR (OK), around Boulder (CO), and in Comanche NG (CO). Occurs in grassy fields and forest clearings e. to w.-cen. LA, where it has been recorded in Kisatchie NF.

LONG-TAILED POCKET MOUSE (*Chaetodipus formosus*). Locally common in dry areas with sparse shrubs, from se. CA to Great Salt Lake (UT), usually on rocky or gravelly soil. Relatively easy to see on warm nights in Joshua Tree NP (CA), in Fish Springs NWR (UT), in Antelope Island SP (UT), along Hwy. 196 (UT), and around Searchlight (NV).

ROCK POCKET MOUSE (*Chaetodipus intermedius*). Locally common in rocky deserts, on canyon bottoms, and around boulder piles in AZ and NM. Occurs in the foothills of Gallinas, Sacramento, and Magdalena Mts. and in Prehistoric Trackways NMT (all in NM); common around Tucson (AZ) and along Ruby Rd. north of Nogales (AZ). A beautiful black morph lives on lava flows around Chloride (NM).

SPINY POCKET MOUSE (*Chaetodipus spinatus*). Uncommon in desert washes, on gravel slopes, and in tamarisk groves of se. CA. Often sufficiently tame to be caught by hand (the spines in its fur are too soft to pierce human skin). Occurs in Cibola NWR (AZ/CA), Joshua Tree NP, and Anza-Borrego Desert SP (both in CA).

CALIFORNIA POCKET MOUSE (*Chaetodipus californicus*). Inhabits sagebrush and low chaparral in s. CA. Common on dry, chaparral-covered slopes in Santa Cruz and Santa Lucia Mts. (for example, along Cone Peak Trail) and in Santa Monica Mts. NRA. The distinctive San Diego race occurs sw. of I-5 exit 71 near Encinitas. If you hear noise in the chaparral, lie down and look underneath the shrubs; sometimes you'll see a pocket mouse.

SAN DIEGO POCKET MOUSE (*Chaetodipus fallax*). Uncommon and local in areas of hard soil in sw. CA. Look for it in desert flats and canyons around Cabazon, in the southern lowlands of Joshua Tree NP, and in Anza-Borrego Desert SP. The coastal race can be found sw. of I-5 exit 71 near Encinitas.

NELSON'S POCKET MOUSE (*Chaetodipus nelsoni*). Locally common in dry rocky areas of w. TX and se. NM; more active in winter. I've seen it in Tuff Canyon in Big Bend NP (TX) and along the road to Sitting Bull Falls (NM), where on a moonless night up to a dozen could be found with a red flashlight in one hour.

DESERT POCKET MOUSE (*Chaetodipus penicillatus*). Common in sandy desert washes from se. CA to w. TX. Occurs in Joshua Tree NP, Mecca Hills WA, and Anza-Borrego Desert SP (all in CA); near Yuma (AZ) and Florence Junction (AZ). Animals from NM and TX are often considered a separate species, **CHIHUAHUAN POCKET MOUSE** (*C. eremicus*). Look for this darker form in Big Bend Ranch SP and along the Rio Grande in Big Bend NP (both in TX), as well as in Animas Valley (NM).

BAILEY'S POCKET MOUSE (*Chaetodipus baileyi*). Occurs at the foot of rocky slopes in hot deserts of s. AZ. Reportedly common around Portal, in the area below Madera Canyon, on the lands of Tohono O'Odham Nation, and in San Pedro Riparian NCA. This species is not supposed to be particularly difficult to find, but for some reason I've seen it only once, near Spud Rock in Saguaro NP.

BAJA POCKET MOUSE (*Chaetodipus rudinoris*). Rare and local at the foot of rocky slopes in sandy deserts of sw. CA; more common after winters with good rainfall. Anza-Borrego Desert SP is a good place to look for it (try the entrance area of Borrego Palm Canyon).

PLAINS POCKET MOUSE (*Perognathus flavescens*). Common and widespread in sandy areas of the Great Plains and locally in the sw. U.S., particularly in regularly overgrazed pastures, prairie dog towns, dunes, and near bison wallows. Look for it in the vicinity of yuccas and chollas in Comanche NG (CO), in the Sandhills of NE, and in White Sands NMT (NM), where a beautiful white morph occurs. Dark morphs can be seen in Sheyenne NG (SD) and in Canyons of the Ancients NMT (CO). The large western race formerly known as **APACHE POCKET MOUSE** (*P. apache*) can be found in Canyonlands NP (UT), around Zuni Pueblo (NM), in Chaco Culture NHP (NM), and around Portal (AZ).

OLIVE-BACKED POCKET MOUSE (*Perognathus fasciatus*). Common in the northern part of the shortgrass prairie zone, particularly in sandy areas and in recently disturbed habitats such as roadside weed patches. An agile runner capable of long leaps, this small rodent can sometimes be identified by its speed as it crosses the road in front of your car. Usually easy to find on moonless nights in places like Grasslands NP (SK), Wind Cave NP (SD), Dinosaur NMT (CO/UT), and Thunder Basin NG (WY).

SILKY POCKET MOUSE (*Perognathus flavus*). Locally abundant in the sw. U.S. in prairies and foothills with sparse vegetation, often on bare sand or rock. Common in prairie canyons of Comanche NG (CO), in Cimarron NG (KS), in Lower Rio Grande Valley NWR (TX), and in many locations in NM, such as at Lordsburg Playas, south of Placitas, around Zuni Pueblo, in Sevilleta NWR, and in Chaco Culture NHP. The

When found, Silky Pocket Mouse will often try to hide in plain sight rather than run away.

distinctive southwestern form can be found in Organ Pipe Cactus NMT and around Portal (both in AZ).

MERRIAM'S POCKET MOUSE (*Perognathus merriami*). Occurs in sparsely vegetated scrub deserts of TX and e. NM; common in Big Bend NP (TX) and locally around Clovis (NM). Unlike most other pocket mice, it is usually shy and difficult to see; covering your flashlight with red film and wearing soft shoes might help.

LITTLE POCKET MOUSE (*Perognathus longimembris*). Locally common from spring to late summer in arid plains and foothills of the sw. U.S., mostly in dry grasslands on gravel or rock. Too tiny to trigger many rodent traps, it is often recorded as rare by biologists conducting trapping surveys despite being really common. Usually stealthy, quiet, and difficult to see well. Common along Munz Ranch Rd. near Lake Hughes (CA), around Mono Lake (CA), in Sheldon NWR (NV), and in Desert NWR (NV); in some years also in Snow Canyon SP (UT) and Vermilion Cliffs NMT (AZ). The isolated coastal race (**PACIFIC POCKET MOUSE**), which occurs in coastal chaparral of s. CA (for example, at Dana Point Headlands and in Border Field SP), is listed as endangered in the U.S.

SAN JOAQUIN POCKET MOUSE (*Perognathus inornatus*). Common at scattered locations in dry grasslands of CA, mostly in the s. Central Valley. Often very tame and easy to approach, particularly with red light. Easy to see in most years in the southern part of Carrizo Plain NMT. Present but more difficult to find along Little Panoche Rd. and in Vic Fazio Yolo WLA.

Little Pocket Mouse is sometimes so tame that it can be hand-fed.

ARIZONA POCKET MOUSE (*Perognathus amplus*). Common in hot lowland deserts of AZ and s. NV. Relatively easy to find on moonless nights in overgrown desert washes in Organ Pipe Cactus NMT, and in Las Cienegas NCA; widespread on the lands of Tohono O'Odham Nation (all in AZ).

GREAT BASIN POCKET MOUSE (*Perognathus parvus*). Common in areas with sparse vegetation in sagebrush plains and along desert streams in the interior West. Easy to find, for example, in Hanford Reach NMT (WA), Malheur NWR (OR), in City of Rocks NR (ID), in Stillwater NWR (NV), around Mono Lake (CA), and at Steens Mt. (OR). Rare in Canada, where it occurs in arid parts of the Okanagan Valley (BC). The distinctive yellow-eared form can be seen (with much effort) on the southeastern side of Tehachapi Pass (CA).

WHITE-EARED POCKET MOUSE (*Perognathus alticolus*). Very rare and local on sandy slopes in the mountains of s. CA. Difficult to trap, but can be found by spotlighting in appropriate habitat. Look for it in clearings with sparse grass in pine forests on the southeastern side of Tehachapi Pass and around Lake Hughes. Some data indicate that it is a subspecies of Great Basin Pocket Mouse, but on Tehachapi Pass these two taxa occur very close to each other. If you are planning to look for rodents at Tehachapi Pass, it's a good idea to get in contact with biologists working in the area, since any observations you collect might be of scientific interest.

DARK KANGAROO MOUSE (*Microdipodops megacephalus*). This impossibly cute, toylike creature is locally common in sagebrush deserts of the Great Basin. Occasionally seen while driving across NV along Hwy. 50 or Hwy. 6 on moonless nights. Look for it after sunset east of the entrance to Great Basin NP (NV), around Beatty (NV), in sand dunes near Alkali Lake (OR), and around Mono Lake (CA).

PALE KANGAROO MOUSE (*Microdipodops pallidus*). This bouncing ball of a rodent is said to be locally common in some years in sandy deserts of NV and e. CA. I've seen it only once, near Sand Mt. (NV), after following its tracks for a couple of hours at night. Reportedly occurs in Deep Springs Valley (CA).

ORD'S KANGAROO RAT (*Dipodomys ordii*). Abundant in arid habitats throughout the interior West, particularly in sandy areas; usually the most common kangaroo rat within its range. Frequently seen on night drives and walks in all NGs of the shortgrass prairie zone, particularly in Cimarron NG (KS) and Thunder Basin NG (WY). Common also in City of Rocks NR (ID), Great Sand Dunes NP (CO), Rabbit Valley (CO), Aubrey Valley (AZ), Nebraska NF (NE), Bruneau Dunes SP (ID), Chaco Culture NHP (NM), the Placitas area (NM), Grand Staircase-Escalante NMT (UT), Beaver Dam Wash NCA (UT), the vicinity of Falcon SP (TX), and numerous other protected areas; also around Mono Lake (CA) and along Hwy. 50 and Hwy. 6 (NV). The grizzled-gray northern subspecies can best be found in Great Sandhills (SK) in October and early November, after it has molted into lush winter fur.

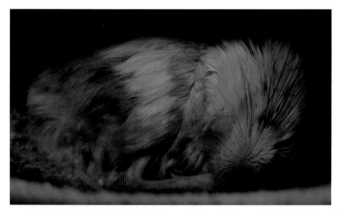

Ord's Kangaroo Rat in its burrow. This species prefers sandy soil, and is often easy to track.

GULF COAST KANGAROO RAT (*Dipodomys compactus*). Locally common in sandy, sparsely vegetated habitats of s. TX. The almost-white barrier islands race is abundant in sand dunes in Padre Island NSS, where in good years dozens can be seen in one night. Mainland individuals are darker-colored; look for them around Riviera Beach. Often very tame and easy to approach—best with red light.

DESERT KANGAROO RAT (*Dipodomys deserti*). This large, handsome, light-colored rodent is locally abundant in the hottest lowlands of the sw. U.S. Good places to look for it include Snow Canyon SP (UT); Amargosa Valley, the Hidden Cave area, and the Pyramid Lake area (all in NV); the edges of dune fields in Death Valley NP (CA); desert washes in Joshua Tree NP (CA); the Colorado River Valley along the CA/AZ state line; and Kofa NWR (AZ).

BANNER-TAILED KANGAROO RAT (*Dipodomys spectabilis*). This impressive creature is locally common in the sw. U.S. in desert grasslands with hard soil and lots of shrubs, where it builds huge mounds of sand with up to a dozen openings, surrounded by a network of runways. Usually easy to find around Rodeo, in Bosque del Apache NWR, Animas Valley, and Chaco Culture NHP (all in NM); around San Bernardino NWR (AZ); and in Davis Mts. SP (TX).

TEXAS KANGAROO RAT (*Dipodomys elator*). Very rare and local in arid mesquite grasslands with clay soils of n.-cen. TX, where it builds burrows at the bases of mesquite trees. Can be found with considerable effort by driving on moonless nights along backcountry roads between Santa Rosa Lake and Copper Breaks SPs.

MERRIAM'S KANGAROO RAT (*Dipodomys merriami*). Common in open deserts from nw. NV to w. TX. It is less dependent on sandy soils than other kangaroo rats, and can often be encountered on hard soils and even rocky terrain. Good places to look for it include the Mono Lake area (CA), Anza-Borrego Desert SP (CA), Joshua Tree NP (CA), Black Rock Desert (NV), Snow Canyon SP (UT), Big Bend NP (TX), the foothills south of Placitas (NM), Animas Valley (NM), and all desert parks of s. AZ. The small, dark-colored subspecies from San Bernardino Mts. (CA) is listed as endangered in the U.S.; it is known to occur along Lytle Creek, but I've never been able to find it there.

CHISEL-TOOTHED KANGAROO RAT (*Dipodomys microps*). Common in sagebrush deserts and sparsely wooded foothills of the Great Basin. Easy to see, for example, around Mono Lake (CA), along the access road to Great Basin NP (NV), in Antelope Island SP (UT), in Fish Springs NWR (UT), and in sand dunes near Alkali Lake (OR).

Frequently seen while crossing NV along Hwy. 50 or Hwy. 6. Distinctive races occur in Stubbe Springs and Skull Rock areas of Joshua Tree NP (CA), in w. and s. portions of Vermilion Cliffs NMT (AZ), and on islands in Great Salt Lake (UT).

FRESNO KANGAROO RAT (*Dipodomys nitratoides*). Also known as **SAN JOAQUIN KANGAROO RAT**, this handsome rodent is rare and local in arid plains and sandy hills of the s. Central Valley of CA. Easy to see (except after dry winters) in Carrizo Plain NMT. Occurs at Kettleman Hills, in Pixley NWR, in Alkali Sink Ecological Reserve, in Mendota WMA, and along the northeasternmost part of Little Panoche Rd. Two of its three subspecies are listed as endangered in the U.S.

CALIFORNIA KANGAROO RAT (*Dipodomys californicus*). Local and difficult to find in dry habitats of n. CA and s.-cen. OR. It was common in the late 1990s around Lake Beryessa (CA), but the habitat there has been mostly destroyed by development; nearby Stebbins Cold Canyon Reserve still has a small population. Reportedly occurs in Clear Lake SP (CA), on dry low-elevation slopes in Trinity Alps (CA) and the Kalmiopsis Wilderness (OR), in vacant lots around Altamont (OR), and in dry meadows in a part of Cascade-Siskiyou NMT called Mariposa Lily Botanical Area, located on the western side of I-5 about a mile north of the OR/CA state line.

GIANT KANGAROO RAT (*Dipodomys ingens*). Very local in the southwestern part of the Central Valley (CA). Usually easy to see by slowly driving late at night around the southern part of Carrizo Plain NMT, but can be hard to find in years with poor rainfall in winter and spring. Also occurs along Little Panoche Rd., particularly in Panoche Hills Management Area (signposted just n. of Mercey Hot Springs). Listed as endangered in the U.S.

Although Giant Kangaroo Rat is now surviving in only a few small areas of California, it can be easy to see on moonless nights if you know where to look. Some individuals are surprisingly unafraid of people and cars, so please drive very carefully when searching for them.

AGILE KANGAROO RAT (*Dipodomys agilis*). Uncommon and local in upland deserts of s. CA. I've seen it in sandy areas with lots of shrubs in and around Antelope Valley California Poppy Reserve, and along Munz Ranch Rd. near Lake Hughes. Reportedly common in dry canyons in Santa Monica Mts. NRA.

DULZURA KANGAROO RAT (*Dipodomys simulans*). Uncommon and local in sw. CA. Occurs in sandy washes in Mecca Hills WA and Anza-Borrego Desert SP. Also common in sparse chaparral sw. of I-5 exit 71 near Encinitas. If you attempt to look for it there, bring your ID: very suspicious border patrol. Abundant in Mexico around Guerrero Negro (Baja Sur). Possibly a subspecies of Agile Kangaroo Rat.

NARROW-FACED KANGAROO RAT (*Dipodomys venustus*). Rare and local on chaparral-covered slopes with sandy soils in w.-cen. CA. Occurs along the access road to the Cone Peak Trail in the Santa Lucia Mts., on dry mountaintops in the Santa Cruz Mts., and at low elevations in Mt. Diablo SP. The rare Gabilan Mts. race, formerly considered a separate species, **BIG-EARED KANGAROO RAT** (*D. elephantinus*), is somewhat common only in Pinnacles NP, in unburned chamise chaparral with lots of bare ground (try Bear Gulch Reservoir area).

HEERMANN'S KANGAROO RAT (*Dipodomys heermanni*). Occurs in dry habitats with sparse ground cover in cen. CA. Usually common in Carrizo Plain NMT; present along Little Panoche Rd., around La Panza, in Pinnacles NP (CA), and in the eastern part of Mt. Diablo SP. Its numbers fluctuate a lot, but usually you can see at least a few in one night of driving. The isolated subspecies inhabiting sand dunes around Morro Bay is listed as endangered in the U.S.

PANAMINT KANGAROO RAT (*Dipodomys panamintinus*). Locally common in dry foothills with scattered shrubs and/or Joshua trees along the eastern side of the Sierra Nevada. Occurs near Mono Lake, along the road to Dante's View in Death Valley NP, on the southeastern side of Tehachapi Pass, in Deep Springs Valley (all in CA), and around Searchlight (NV).

STEPHENS'S KANGAROO RAT (*Dipodomys stephensi*). The most difficult N. American kangaroo rat to see, it is very rare and local in sw. CA. Look for it on moonless nights on sparsely vegetated slopes of San Jacinto WLA and Lake Perris SRA. The largest surviving population lives on private land (at Warner's Ranch near Lake Henshaw); I don't know if it's possible to obtain access permission. Listed as endangered in the U.S.

MICE, RATS, AND JUMPING MICE

These rodents are usually so easy to trap that observational studies of their behavior in the wild are largely neglected. However, they are generally easy to find by spotlighting, especially the more noisy species such as woodrats and jumping mice. Most of them are almost strictly nocturnal and are easier to see on dark nights, but cotton rats, Northern Pygmy Mouse, and *Rattus* rats are often active during the day. Jumping mice hibernate (go dormant in winter), while some desert *Peromyscus* species estivate (go dormant in summer). In arid parts of the West, the population densities of many species fluctuate depending on rainfall, so the best time to look for them is in early fall after a summer with good monsoon rains.

Most species feed on seeds, insects, fungi, and other high-value food, so they have to travel far from their nests and burrows to find these morsels. They are very agile and alert. Jumping mice are particularly fast; these relatives of jerboas often prefer hopping to running, and can make leaps up to 10 ft. (3 m) long. Rice and cotton rats are more volelike in their behavior: they eat more grass and sedges and so don't have to move around that much. Instead, they have small, cozy home ranges with well-maintained runway networks, backup nests, feeding platforms, and other comforts. Mice seldom make their own runways, but if you are looking for voles by watching vole runways, expect to see mice using them as well. Another good method is to watch woodrat houses; they can be used by both woodrats and mice.

Most species are good climbers and can be partially arboreal. *Peromyscus* mice often nest in bird nest boxes, especially in winter. Golden Mouse and harvest mice build ball-shaped nests on the ground or in low vegetation, or sometimes occupy old bird nests and build a dome-shaped roof over them. Woodrats can often be located by the rattling sound they make with their feet or teeth, while grasshopper mice and some harvest mice can reveal themselves with their high-pitched calls.

Some species, particularly Deer Mouse and its western relatives, regularly carry the deadly Sin Nombre hantavirus, and most species can carry plague and tularemia, so all should be handled with great care.

MARSH RICE RAT (*Oryzomys palustris*). Common in marshes and meadows from TX to NJ; abundant in many coastal wetlands and rice fields. Makes networks of runways leading to feeding platforms and grass nests. Formerly very common

Marsh Rice Rat adapts very well to habitat modification, and can be abundant in low-lying fields.

in Everglades NP (FL), where it was often seen on roadsides on moonless nights. The population there has crashed following the introductions of Burmese python and Argentine fire ant, but might eventually recover. Better places include Caddo Lake SP (TX), Anahuac NWR (TX), White Lake WCA (LA), Ocmulgee NMT (GA), Okefenokee NWR (GA), Pocosin Lakes NWR (NC), Prime Hook and Bombay Hook NWRs (DE), and Blackwater NWR (MD). The distinctive Lower Keys race, formerly considered a separate species, **SILVER RICE RAT** (*O. argentatus*), occurs in freshwater marshes on Big Torch Key (FL), and can sometimes be seen from Dorn Rd.; it is listed as endangered in the U.S.

COUES'S RICE RAT (*Oryzomys couesi*). Locally common in extreme s. TX in heavily overgrown marshes and meadows. At night it can sometimes be seen swimming across ponds and canals. Look for its distinctive runways and feeding platforms in Bentsen-Rio Grande Valley SP and Laguna Atascosa NWR.

EASTERN HARVEST MOUSE (*Reithrodontomys humulis*). Locally common in the se. U.S., where it occurs in fallow fields, vacant lots, meadows, marshes, and wet prairies. This tiny rodent tends to be inconspicuous and difficult to see or trap; I've seen it only a few times in many nights of searching for it. It is present in the newly created Everglades Headwaters NWR (FL), in Sabine NWR (LA), in and around Bond Swamp NWR (GA), in Carolina Sandhills NWR (SC), in Whites Mill Refuge (TN), in the Cades Cove area of Great Smoky Mts. NP (TN), in Evans Tract (37°48'N, 79°46'W) near Staunton (VA), in Mammoth Cave NP (KY), and in Wayne NF (OH). However, it is not particularly easy to find in any of these places, so if you live anywhere within its range, it's probably better to look for it in good tallgrass habitat around your home than to travel elsewhere.

FULVOUS HARVEST MOUSE (*Reithrodontomys fulvescens*). Rare and local in AZ and Animas Valley (NM); common and widespread from TX to MO and MS in shrubby meadows, forest clearings, woodland edges, and fields. Although mostly nocturnal, it can sometimes be seen at dusk if the weather is unusually cold. Listen for its loud, very high-pitched double call after dark, and look for a tiny ball of fur climbing through dense grass. Relatively common in Attwater Prairie Chicken NWR (TX), in Caddo Lake SP (TX), in St. Catherine Creek NWR (MS), and particularly in Duralde Cajun Prairie near Eunice (LA).

PLAINS HARVEST MOUSE (*Reithrodontomys montanus*). Widespread over much of the Great Plains; local in intermontane valleys of CO, NM, and se. AZ. Inhabits shortgrass prairies, meadows, fields, and overgrown pastures, but is often missing from seemingly appropriate habitat. Look for its nests built just aboveground in low-elevation portions of Rocky Mt. NP (CO), in Crescent Lake NWR (NE), in Wichita Mts. WR (OK), and along Hwy. 80 north of Douglas (AZ).

WESTERN HARVEST MOUSE (*Reithrodontomys megalotis*). Widespread in the West and parts of the Midwest, mostly in fallow fields, meadows, forest clearings, marshes, planted cereals, and mixed-grass prairies. In CA it also occurs in oak savannas, arid grasslands, gardens, and sometimes in houses. Very rare in Canada. Occasionally can be found after sunset by its high-pitched trilling call. Highly variable in appearance, it is usually brown in coastal areas (east to Sierra Nevada crest) and grayish in the arid interior. Common in appropriate habitats in Humboldt Lagoons SP, in the Santa Cruz Mts., in Stebbins Cold Canyon Reserve, and in Elkhorn Slough NR (all in CA), in Hanford Reach NMT (WA), around Boulder (CO), in City of Rocks NR (ID), near Zuni Pueblo (NM), in Fish Springs NWR (UT), and in Tallgrass Prairie NPR (KS). Often seen at dusk and even during the day around the bird feeders at the visitor center in Patagonia Lake SP (AZ). Distinctive races inhabit coastal shrublands and wetlands on Santa Cruz and Catalina Is. (CA). The northernmost populations fluctuate annually, with peak numbers in late fall, so October and November are the best time to look for this mouse in Okanagan Valley (BC) and Suffield NWR (AB).

SALT-MARSH HARVEST MOUSE (*Reithrodontomys raviventris*). Endemic to the San Francisco Bay Area (CA), where it occurs in well-vegetated salt marshes and sometimes in adjacent grasslands. Look for this tiny rodent during winter high tides (see details in California and Nevada section of Part II) at Hayward Regional Shoreline, in Don Edwards San Francisco Bay NWR, and in Palo Alto Baylands Nature PR. The white-bellied northern subspecies occurs in San Pablo Bay NWR (try the loop part of Lower Tubbs Island Trail) and in Grizzly Island WLA, but is more difficult to see. Listed as endangered in the U.S.

GOLDEN MOUSE (*Ochrotomys nuttalli*). This semiarboreal species is widespread, but not particularly common, in the se. U.S., where it inhabits vine tangles, wood edges, tall shrubs, forests with dense undergrowth, and overgrown rocky slopes. Often feeds up to 17 ft. (5 m) aboveground. It has highly reflective golden fur and looks like an exquisite jewelry piece when caught in a flashlight beam. Look for it in Torreya SP (FL) and around Mead's Quarry in Knoxville (TN) during the second half of the night. Common in appropriate habitat in Bond Swamp NWR (GA), Cumberland Gap NHP (TN/KY/VA), Bogue Chitto NWR (LA), and Big Thicket NPR (TX).

FLORIDA MOUSE (*Peromyscus floridanus*). Uncommon and local in parts of FL in open, arid habitats; it's more common in FL sand scrub patches, for example, in Jonathan Dickinson SP, in Wekiwa Springs SP, and in the Big Scrub area of Ocala NF. At Archbold Biological Station, look for its trails crossing sandy roads in the area across the railroad from the headquarters. A few hours of predawn watch at such a trail is a reliable way to see this mouse, but expect it to cross the road *very* quickly—you'll barely have enough time to switch on your flashlight. Occasionally it can be seen at night by looking into gopher tortoise burrows.

DEER MOUSE (*Peromyscus maniculatus*). This is the world's most versatile rodent; it occurs in almost every habitat in N. America, from the deserts of AZ to the tundra near Churchill (MB), high mountains, and isolated houses. In se. U.S. it is mostly confined to high elevations in the Appalachians. More than a hundred subspecies have been described, but they fall into six distinctive types (probably different species). The eastern ("woodland") form can be seen, for example, at the summit of Mt. Mitchell (NC), along the road to Clingmans Dome in

Great Smoky Mts. NP (TN/NC), in the Adirondack and Catskill Mts. (NY), anywhere in New England, at Linwood Springs Research Station (WI), in Algonquin PP (ON), in La Mauricie NP (QC), and in Kouchibouguac NP (NB). The "prairie" form is the most abundant rodent of the Great Plains, particularly in and around prairie dog towns. Look for it in any NG or NP in the prairie zone, and around Boulder (CO). The "desert" race is easy to see almost everywhere in NM and AZ; it often lives in houses, for example, on the lands of the Navajo Nation (AZ/NM). "Mountain" races are the most abundant small rodents, for example, in the Sandia Mts. (NM), in the Uinta Mts. (UT), in City of Rocks NR (ID), at Spanish Peaks (CO), and in Rocky Mt. NP (CO). The "bo-

Deer Mouse is charming and interesting to watch, but it can carry the deadly Sin Nombre virus, so if these rodents settle in your home or garage, it's better to trap them out.

real" type is common, among other places, in Yellowstone NP (WY/MT/ID) and in Banff NP (AB). It is somewhat less common on the Pacific Coast, where it occurs in all types of forest, and sometimes in caves and lava tubes such as Ape Cave in Mt. St. Helens NMT (WA). The distinctive Channel Is. races can be seen on San Miguel, Santa Cruz, Anacapa, and Catalina Is. (CA).

deep canyons, and windfall areas. Often seen at night on vertical rock faces, near tree hollows, and in brush piles. Can be abundant after good monsoon rains in piñon-juniper and oak woodlands of NV, NM, AZ, UT, and CO. Sometimes occurs in abandoned houses. Common along Mineral King Rd., in Salt Point SP, Stebbins Cold Canyon Reserve, and Calaveras Big Trees SP (all in CA); in prairie canyons of Comanche NG (CO); in Mesa Verde NP (CO); in the foothills of the Sandia Mts. (NM); in Chaco Culture NHP (NM); along both rims of the Grand Canyon (AZ); in Kodachrome Basin SP (UT); and in the Wichita Mts. WR (OK).

TEXAS MOUSE (*Peromyscus attwateri*). Locally common from cen. TX to sw. MO, mostly in open woodlands on rocky slopes. Often feeds in small trees, in shrubs, and on rock surfaces. Relatively tame and easy to spotlight, for example, near Caverns of Sonora (TX), in Copper Breaks SP (TX), and in the Ouachita Mts. (OK/AR).

WHITE-ANKLED MOUSE (*Peromyscus pectoralis*). Uncommon in a variety of arid habitats of w. TX, usually with rocks, shrubs, small trees, and a layer of dry leaves. Occurs in Big Bend NP, for example, near Chisos Basin Lodge; common in wet years in rocky areas along Rd. 2627 east of the NP and in Copper Breaks SP (all in TX).

NORTHERN ROCK MOUSE (*Peromyscus nasutus*). Uncommon and local in the sw. U.S., mostly in piñon-juniper-oak woodlands on rocky slopes, in canyons, and on old lava fields. Look for it in Carlsbad Caverns NP, El Malpais NMT, Gila Cliff Dwellings NMT, and the Sandia Mts. (all in NM), as well as in Canyons of the Ancients NMT (CO) and in Valley of the Gods (UT).

OSGOOD'S MOUSE (*Peromyscus gratus*). Occurs in sw. NM and adjacent parts of AZ on rocky slopes with shrubs and small trees; it's usually difficult to find even in good habitat. Present in Granite Gap (NM), around Gila Cliff Dwellings NMT (NM), and on old lava flows near Paramore Crater (AZ). Easier to see in Mexico, where it's abundant, for example, in dry coniferous forests above Copper Canyon (Chihuahua).

PIÑON MOUSE (*Peromyscus truei*). Widespread but uncommon in arid parts of the sw. U.S., where it prefers rocky slopes with large piñons and junipers, but occurs also in open pine woods and tall shrubs. In CA and OR it also lives in redwood groves, chaparral, and oak savannas. Often climbs trees, where it hides from flashlight beam behind branches; look for its huge Mickey Mouse–like ears sticking out. Generally tame and can sometimes be approached very closely, or even touched. Look for it in Pinnacles NP, in Stebbins Cold Canyon Reserve, along the eastern approaches to Tioga and Sonora Passes, along Mineral King Rd., around Mt. Whitney Fish Hatchery, and in the Santa Cruz Mts. (all in CA); in prairie canyons of Comanche NG (CO); in Agua Fria NMT (AZ); in the foothills of the Sandia Mts. (NM); and around Zuni Pueblo (NM). The distinctive race formerly known as **COMANCHE MOUSE** (*P. comanche*) is common in Palo Duro Canyon SP (TX).

CALIFORNIA MOUSE (*Peromyscus californicus*). Common in CA south of Sacramento River Delta in all kinds of shrublands and woodlands, but mostly in chaparral and dry foothill forests. Frequently lives in woodrat nests, particularly in the Sierra Nevada; that habit earned it an old name, **"PARASITIC MOUSE."** Often very tame and not afraid of a flashlight. Easy to see in Los Padres NF, along the lower part of Mineral King Rd., along Cone Peak Trail and Nacimiento-Ferguson Rd. in the Santa Lucia Mts., in the Santa Cruz Mts., and in Fort Ord NMT.

NORTHERN GRASSHOPPER MOUSE (*Onychomys leucogaster*). This chubby predator is common and widespread in the interior West, particularly in dry open areas, but also in arid woodlands. Easy to see on dark nights in the shortgrass prairie zone, when it patrols the edges of prairie dog towns. Look for it in Grasslands NP (SK), Black Canyon of the Gunnison NP (NM), Aubrey Valley (AZ), City of Rocks NR (ID), Thunder Basin NG (WY), Comanche NG (CO), Cimarron NG (KS), around Boulder (CO), and near Zuni Pueblo (NM). Often abundant in cornfields in ND and MT and in sagebrush flats in Hanford Reach NMT (WA).

Grasshopper mice are fierce predators, capable of a nasty bite. But they love having their belly tickled.

MEARNS'S GRASSHOPPER MOUSE (*Onychomys arenicola*). Uncommon in parts of the sw. U.S. in lowland deserts with creosote bush and mesquite. Look for it in Carlsbad Caverns NP, Sevilleta NWR, Prehistoric Trackways NMT (all in NM), and Big Bend NP (TX). As with all grasshopper mice, it can sometimes be located at night by its high-pitched territorial call, which sounds a bit like a whistling kettle.

SOUTHERN GRASSHOPPER MOUSE (*Onychomys torridus*). Generally uncommon in parts of the sw. U.S. in deserts and arid grasslands; often lives in ground squirrel colonies and hunts in surrounding desert. Easy to find in most years around Onyx (CA), in Joshua Tree NP (CA), along Rte. 66 between Seligman and Aubrey Valley (AZ), and in Kofa NWR (AZ). Distinctive races occur in Carrizo Plain NMT and sw. of I-5 exit 71 near Encinitas (both in CA).

NORTHERN PYGMY MOUSE (*Baiomys taylori*). Uncommon and local but increasing in parts of the sw. U.S. Isolated populations are scattered through open areas with good ground cover, from cholla patches in deserts to open woodlands. Partly diurnal, and can sometimes be active on the surface even on hot days, but seldom ventures into the open. Occurs in San Bernardino NWR (AZ), around Rodeo (NM), and in Falcon SP (TX). Relatively common in Aransas and Laguna Atascosa NWRs (TX).

EASTERN WOODRAT (*Neotoma floridana*). Widespread in bottomland forests of the se. U.S.; very local on the Great Plains, where it occurs in canyons and along rocky bluffs. Generally rare and difficult to find. Builds large nests in sheltered places such as hollow trees and logs, abandoned buildings, caves, vine tangles, and armadillo and gopher tortoise burrows. In early spring it wanders far from the nest, and can occasionally be seen during late-night drives on quiet forest roads. Often climbs trees, particularly acorn-bearing oaks. Look for it in dense woods of Croatan NF (NC), Wichita Mts. WR (OK), and Quivira NWR (KS); along limestone cliffs in Mammoth Cave NP (KY), Big South Fork NRA (TN/KY), Pickett SP (TN), and Linville Gorge (NC); in swampy hardwood forests in Big Thicket NPR (TX), Atchafalaya NWR (LA), and Bond Swamp NWR (GA); and in areas with lots of palmetto in Highlands Hammock SP (FL). The distinctive Key Largo race (listed as endangered in the U.S.) can be seen with some luck in Dagny Johnson Key Largo Hammock Botanical SP (FL).

APPALACHIAN WOODRAT (*Neotoma magister*). Also called **ALLEGHENY WOODRAT**, this species has been recently split from Eastern Woodrat. It is rare and local in hills and mountains of the Appalachian region, and largely extinct on surrounding plains. Lives in dense woods, particularly with red cedar; builds cup-shaped nests in rock crevices, cliff niches, caves, and talus slopes. Still relatively common only in WV and parts of VA, for example, in Shenandoah NP (VA) and in Monongahela NF (WV). Present in Great Smoky Mts. NP (TN/NC), where I've seen it a couple times while driving late at night along the access road to Cades Cove.

White-throated Woodrat's nests provide shelter for mice, shrews, and many other small animals.

WHITE-THROATED WOODRAT (*Neotoma albigula*). Common in desert shrub, arid woodlands, and on rocky slopes and cactus flats of the sw. U.S. Builds large stick houses, usually under cacti or in rock crevices, and often covers them with jumping cholla twigs. Common, for example, in Havasu NWR, Organ Pipe Cactus NMT, and Agua Fria NMT (all in AZ); also in Chaco Culture NHP, around Zuni Pueblo, and in the western part of Sevilleta NWR (all in NM).

WHITE-TOOTHED WOODRAT (*Neotoma leucodon*). Also known as **EASTERN WHITE-THROATED WOODRAT**, it is locally common in arid woodlands, overgrazed prairies, rocky areas, and deserts with lots of cacti in NM (except the western part) and w. TX. Occurs, for example, around Placitas (NM) and in Big Bend NP (TX). Dark-colored woodrats live on lava flows in El Malpais NMT (NM). Probably a subspecies of White-throated Woodrat.

DESERT WOODRAT (*Neotoma lepida*). Common and widespread in the sw. U.S. in deserts and sometimes in arid woodlands, usually in areas with lots of cacti and yuccas. Houses are often built in cholla clumps or in rock crevices under large cacti. Common in parts of Death Valley NP, in Joshua Tree NP, in Carrizo Plain NMT, in Pinnacles NP (all in CA), and in Cathedral Gorge SP (NV). The distinctive coastal race, often considered a separate species **INTERMEDIATE WOODRAT** (*N. intermedia*), is common in dry, sagebrush-covered hills above Malibu (CA); tame individuals can be seen just after sunset at the summit of Cone Peak (CA).

ARIZONA WOODRAT (*Neotoma devia*). A little-known species with uncertain distribution, until recently considered a subspecies of Desert Woodrat. Occurs in deserts with cacti and yuccas, often at the base of rocky cliffs; builds small houses in boulder piles and shady canyons. Relatively common in Kofa NWR and Lake Mead NRA (both in AZ).

MEXICAN WOODRAT (*Neotoma mexicana*). Common in parts of the sw. U.S. in dry, open coniferous woodlands on rocky slopes; very rare in lowland deserts. Builds small nests in cliff crevices and caves. Occurs around Colossal Cave (AZ) and Boulder (CO); also on the western slope of the Sandia Mts. (NM) and in Chaco Culture NHP (NM). Often lives in abandoned buildings along the foothills of the Front Range (CO) and in adjacent prairie areas.

Mexican Woodrat is widespread in arid mountains of the Southwest.

SOUTHERN PLAINS WOODRAT (*Neotoma micropus*). Locally common in the s. Great Plains and much of NM and TX. Lives in all kinds of arid habitats, but is most common on overgrazed pastures, on rocky slopes, in shrub patches, in streamside thickets, and in abandoned homesteads. Builds large houses, usually under a shrub, large cactus, or dilapidated wooden building. Often easy to find in Santa Ana and Laguna Atascosa NWRs (both in TX), where it visits bird feeders at dusk; also in prairie canyons of Comanche NG (CO), around Placitas (NM), and in upland portions of Sevilleta NWR (NM). Possibly a subspecies of Eastern Woodrat.

DUSKY-FOOTED WOODRAT (*Neotoma fuscipes*). Common in dense woodland and sometimes in tall chaparral in parts of CA and OR. Builds large houses on the ground or in trees. Very noisy and conspicuous, and often tooth-chatters in addition to foot-drumming, but can be difficult to see well in dense foliage. Abundant in oak groves in the hills around San Francisco Bay and along the western side of the Central Valley. Look for it in Mt. Diablo SP, in Stebbins Cold Canyon Reserve, in the Santa Cruz Mts. (all in CA), and in Alfred A. Loeb SP (OR). The distinctive riparian race is now practically limited to riverside thickets in Caswell Memorial SP (CA); it is very rare and listed as endangered in the U.S.

LARGE-EARED WOODRAT (*Neotoma macrotis*). Recently recognized as a species separate from Dusky-footed Woodrat, this handsome rodent is common in woodlands with dense undergrowth and in tall chaparral of the Sierra Nevada and coastal s. CA. Builds large houses on the ground or in trees. Noisy, conspicuous, and rather tame. Easy to see around Cone Peak Trailhead, near Chumash Painted Cave SHP, and at lower elevations in Sequoia NP (all in CA).

BUSHY-TAILED WOODRAT (*Neotoma cinerea*). Very widespread in the West, north to YT. Occurs in coniferous forests, juniper woodlands, and rocky canyons, less often in other habitats, usually along streams or on rocky slopes. Builds houses high aboveground or in cool sheltered places such as abandoned houses, mines, and caves. Generally shy and inconspicuous. Partly diurnal in summer in the North. Good places to look for it include Sequoia NP (CA), China Flat Campground (CA), Hells Canyon NRA (OR/ID), Oregon Caves NMT (OR), North Cascades NP (WA), Silver Creek Preserve (ID), City of Rocks NR (ID), the Magdalena Mts. (NM), the North Rim of the Grand Canyon (AZ), Yellowstone NP (WY), Capitol Reef and Canyonlands NPs (UT), Yoho NP (BC), and Muncho Lake PP (BC).

common in HI; abundant and easy to see in Koke`e SP on Kauai, as well as in low-elevation parts of Haleakala NP and in sugar cane fields on Maui. The most common land mammal in Bermuda.

POLYNESIAN RAT (*Rattus exulans*). Introduced to HI by the Polynesians; now rare thanks to competition with Black Rat. Relatively shy and inconspicuous, but can be active in broad daylight in shady montane forests. Look for it along Alakái Swamp Trail in Koke`e SP on Kauai, in Hawaii Volcanoes NP on the Big Island, in Ahupua`a O Kahana SP on Oahu, and in sugar cane fields on Maui.

HOUSE MOUSE (*Mus domesticus*). Recently split from *M. musculus* (another Eurasian species), this introduced rodent is widespread throughout the U.S., much of Canada, and Bermuda. Abundant in small towns and agricultural areas; local and limited to human dwellings in n. Canada and AK. Often easy to see on zoo grounds. Mostly rare in natural habitats; occurs, for example, in forest clearings in Big Thicket NPR (TX), in meadows around Boulder (CO), along wood margins in Mingo NWR (MO), in brackish marshes around San Francisco Bay (CA), and in prairie remnants in sw. PA. Abundant in cane patches along the lower Rio Grande (TX) and in rice-growing areas of LA. Common in HI, where it can be seen in Koke`e SP on Kauai and in other forested areas. Small populations in towns of Greenland partially originate from medieval introductions by the Vikings.

MEADOW JUMPING MOUSE (*Zapus hudsonius*). Very widespread in s. AK, most of Canada, the e. U.S. and the Midwest; rare and local in the sw. and se. U.S. Common in most years in forests, clearings, meadows, marshes, grasslands, and brush patches. Noisy and conspicuous; easy to find by spotlighting or at dusk when nights are short, particularly in areas of high density such as in Wrangell-St. Elias NP (AK), around Hay River (NT), in Fundy NP (NB), in Moosehorn and Sunkhaze Meadows NWRs (ME), in Lonsdale Marsh (RI), in Great Swamp NWR (NJ), in the Catskill and Adirondack Mts. (NY), and in Moshannon SF (PA). Common in many tallgrass prairie reserves, such as Kalsow Prairie SPR (IA), Lostwood NWR (ND), Pipestone NMT (MN), and Midewin National Tallgrass Prairie (IL). Shortgrass prairie races are very difficult to find; they are known to be present in riparian meadows around Boulder (CO), on the grounds of the U.S. Air Force Academy near Colorado Springs (CO), in Rocky Flats NWR (CO), in Bosque del Apache NWR (NM), and in Buffalo Gap NG (SD). The subspecies living in CO and WY (**PREBLE'S MEADOW JUMPING MOUSE**) is listed as threatened in the U.S., and the subspecies living in NM might soon follow.

WESTERN JUMPING MOUSE (*Zapus princeps*). Locally common in meadows, lakeshore marshes, wet thickets, and riparian woodlands in the w. U.S. and w. Canada north to YT. Somewhat less noisy than other jumping mice, it is still relatively easy to find, particularly in the northern part of the range, where it becomes partially diurnal in June and July. Look for it in City of Rocks NR (ID), in high-elevation meadows in Yosemite NP (CA), near McCloud Falls (CA), in Bowman Lake Campground in Glacier NP (MT), in Clark Salyer NWR (ND), near Teslin Lake (YT), and in Glacier, Mt. Revelstoke, Yoho, and Kootenay NPs (all in BC). Recent studies have shown that Sierra Nevada populations might actually belong to Pacific Jumping Mouse.

PACIFIC JUMPING MOUSE (*Zapus trinotatus*). Common in the Pacific Northwest, particularly in moist coastal meadows, in alder groves, along streams in coniferous forests, and on densely vegetated lakeshores. Easier to find by spotlighting than any other rodent within its range, particularly in places with lots of Sewellel burrows. Reliable places include coastal sections of Olympic NP (WA) and Redwood NP (CA), as well as the lower slopes of Mt. Baker (WA) and the vicinity of Bandon (OR). The distinctive southern race occurs in Point Reyes NSS (CA) and Muir Woods NMT (CA).

WOODLAND JUMPING MOUSE (*Napaeozapus insignis*). Very common, conspicuous, and easy to find in most years in the ne. U.S. and e. Canada, but uncommon and local in the Appalachians. Prefers moist, shady coniferous forests; occurs also in hardwood forests, swamps, bogs, and along pond edges and riparian corridors. Seems to be particularly fond of impatiens (jewelweed) patches. Population densities are highly variable, and in some places seem to be low in years when Deer Mouse is more numerous; 2013 was a particularly bad year for this species in much of New England and ON. If disturbed, it will usually make a few erratic leaps and then hide under a leaf or a log. Sometimes can be seen during the day as it hops through the grass. Present, for example, in Colditz Cove SNA (TN), Crane WMA (MA), Acadia NP (ME), the Catskill and Adirondack Mts. (NY), Fundy NP (NB), the Chic-Choc Mts. (QC), Algonquin PP (ON), Laurentides WR (QC), and Kejimkujik NP (NS).

after the first snowfall. Look for it in alpine meadows on Mt. Evans (CO) and Mt. Rainier (WA), in montane valleys around Boulder (CO), in Valles Caldera NPR (NM), in creek-side meadows around Kamloops (BC), around the springs in Ash Meadows NWR (NV), around the end of the marsh boardwalk at the nw. shore of Lake Tahoe (CA), in marshy grasslands in Fish Springs NWR (UT), in grasslands in City of Rocks NR (ID), and in riparian meadows in Yellowstone NP (WY/MT/ ID). The population in Silver Creek Preserve (ID) explodes dramatically once every few years, at which time the voles are very easy to see and attract a lot of predators. Such explosions sometimes occur in other parts of the species' range as well.

CREEPING VOLE (*Microtus oregoni*). Locally common in the Pacific Northwest, mostly in forest clearings, clearcuts, and shortgrass meadows. Largely fossorial and difficult to see, this tiny vole can sometimes be found under sheets of rotten plywood. Since the chances of accidentally finding sheets of rotten plywood in the forests of the Pacific Northwest are pretty slim, consider placing them at good locations yourself and then looking underneath about once every two weeks. Good locations include floodplain meadows along the lower Fraser R. (BC), forest margins in Manning PP (BC), steep grassy slopes in North Cascades NP (WA), stands of dead timber in Mt. St. Helens NMT (WA), forest clearings around Lost Prairie Campground (OR), grassy areas in Nestucca Bay NWR (OR), and meadows in Whis-keytown-Shasta-Trinity NRA (CA).

TOWNSEND'S VOLE (*Microtus townsendii*). A semiaquatic species inhabiting salt marshes, wet floodplain meadows, and swampy riverbanks of the Pacific Northwest. The best places to look for it are the edges of tidal mudflats in river estuaries of coastal OR, WA, Vancouver I. (BC), and the Vancouver City area (BC), where it can be seen swimming between grassy patches during the highest tides. You can also see this vole by following a tractor as it plows a riverside field, or by looking under bales of hay in wet meadows. Occurs in Willapa NWR (WA), in the Newport area (OR), in shoreline meadows at Ross Lake in North Cascades NP (WA), and in Burns Bog on the southern outskirts of Vancouver (BC). A very large, pale-colored subspecies occurs on remote Triangle I. off Vancouver I. (BC). Another island form, small and dark, is found in San Juan Islands NMT (WA).

TUNDRA VOLE (*Microtus oeconomus*). Inhabits wet tundras, muskegs, floodplain meadows, and forest bogs in AK, YT, NT, and nw. BC. Locally abundant in lemming years. This plump vole lives in large, well-designed colonies with extensive networks of vegetation-free runways and huge food stores (its scientific name is derived from the old Russian word for "manager"). These colonies are usually easy to find in appropriate habitat, as long as you don't mind slogging through the tussocks for hours in rubber boots. The best times to try this are late fall, when the bogs freeze over and vole tracks are easy to see on fresh snow, and early spring, when large nests built under snow during the winter become visible. Look around tundra lakes in Kluane NP (YT), in coastal meadows around Yakutat (AK), in peat-swamp spruce forests along the southern part of Dalton Hwy. (AK), in coastal flats on the Kenai Peninsula, and in any open area on Kodiak I. (AK). An even easier way to find this vole is to look in and around abandoned buildings anywhere in rural AK, for example, around Nome. Distinctive races inhabit numerous islands off AK.

PRAIRIE VOLE (*Microtus ochrogaster*). Common from e. Rocky Mts. foothills and the n. Great Plains to w. Appalachian foothills. Inhabits ungrazed prairies, dry meadows, fallow fields, and vacant lots, particularly on sandy soil. Unlike many other prairie rodents, it avoids prairie dog towns. Its runways are well developed but tend to be concealed in dense vegetation. Occurs, for example, in many tallgrass prairie preserves such as Hayden Prairie SPR (IA) and Midewin National Tallgrass Prairie (IL); in fields and lowland meadows around Boulder (CO); and in most WMAs in TN. The small, dark northern race can be seen in areas with lush grass in Theodore Roosevelt NP (ND) and Grasslands NP (SK).

WOODLAND VOLE (*Microtus pinetorum*). Also known as **PINE VOLE**, this rodent is widespread and locally common, but rarely seen, in hardwood forests, small clearings, orchards, and prairie groves of the East and the Midwest. Spends a lot of time underground or under leaf litter, and is difficult to find in most years, in part because its runways are scarce and not very obvious. Even if you locate in by the rustling sound it makes, you might have to wait for hours until you catch a glimpse of it peeking out from underneath dry leaves. Early spring is the best time to look for this small vole, as you can locate its colonies by the remains of under-snow tunnels that become visible as the snow melts. Present in Pinery PP (ON), Crawford Lake CA (ON), Big South Fork NRA (TN), Sequoyah NWR (OK), Piedmont NWR (GA), Mammoth Cave NP (KY), Frozen Head SP (TN), Pinelands NR (NJ), Great Swamp NWR (NJ), Esther Currier WMA (NH), and in woodland lots around Boston (MA).

ROCK VOLE (*Microtus chrotorrhinus*). Uncommon and local in e. Canada; rare in the Appalachians. Prefers moist woods in rocky gulches and below rocky outcrops, often near seeps or small streams with lots of mosses and ferns. The best way to see it is to find a latrine site (usually located in a rock crevice) and watch it; the vole will usually visit it at least once every few hours. Good places include the section of the Appalachian Trail between Hwy. 441 and Clingmans Dome in Great Smoky Mts. NP (NC/TN), rocky outcrops in Shenandoah NP (VA), cave entrances in Allegany SP (NY), the Moose Cave area in Grafton Notch SP (ME), upland areas of Voyageurs NP (MN), shady ravines in Mastigouche WR (QC), canyons in Aiguebelle NP (QC), steep gulches in Cape Breton Highlands NP (NS), and the vicinity of Muskrat Falls near Goose Bay (NL), where the rare, silver-gray Labrador subspecies occurs.

CALIFORNIA VOLE (*Microtus californicus*). Occurs in much of CA and locally in sw. OR, mostly in meadows, marshes, and fields. Abundant and relatively easy to see on foggy mornings in coastal shrublands and drier parts of salt marshes, for example, at Hayward Regional Shoreline, in drier parts of San Pablo Bay NWR (try the access part of Lower Tubbs Island Trail), and particularly in Año Nuevo SP (all in CA). Sometimes occurs in towns, such as Berkeley and Santa Cruz (both in CA). **AMARGOSA VOLE**, the isolated subspecies occurring in Amargosa Valley (NV), is listed as endangered in the U.S.; you can sometimes see these voles on bitterly cold winter mornings in desert wetlands around Tecopa Hot Springs (CA). I highly recommend taking a dip in the hot springs after searching for voles there. Another distinctive race occurs in coastal caparral sw. of I-5 exit 71 near Encinitas (CA).

LONG-TAILED VOLE (*Microtus longicaudus*). Widespread but uncommon in the West; mostly limited to grassy montane areas in NM, AZ, and s. CA; rare and local in AK; more common in WA, OR, and BC. Inhabits riparian and alpine meadows, forests with dense grass, and shrublands. The runways made by this species aren't easily visible, so it's best to look for it after the first snowfall when its tracks are obvious. Relatively easy to find in the highlands of Olympic NP (WA), in forest meadows in Crater Lake NP (OR), and in openings in red fir forests in Yosemite NP (CA). Common also in Kern Canyon (CA), around Waunita Hot Springs (CO), in Valles Caldera NPR (NM), in Kootenay NP (BC), and in moist meadows of Yellowstone NP (WY), City of Rocks NR (ID), Jasper NP (AB), and Yakutat area (AK).

MEXICAN VOLE (*Microtus mexicanus*). Uncommon in high-elevation coniferous forests of the sw. U.S., mostly near old logs in areas with grassy ground cover. Can be found, for example, at Mt. Graham (AZ), along the shores of Baker Lake (AZ), at Mt. Taylor (NM), in Gila Cliff Dwellings NMT (NM), and in high-elevation meadows in Guadalupe Mts. NP (TX). U.S. populations are sometimes considered a separate species from Mexican ones, and called **MOGOLLON VOLE** (*M. mogollonensis*). An isolated race occurring in a few locations in nw. AZ, known as **HUALAPAI MEXICAN VOLE**, is listed as endangered in the U.S.; it's been recently recorded in Pine Peak Canyon in the Hualapai Mts.

SINGING VOLE (*Microtus miurus*). Locally common at the timberline, in streamside willows, and on shrubby tundra of AK and YT. During lemming years this species can be abundant and easy to see as it fre-

Even when abundant, voles of genus Microtus *tend to hide in dense grass, and seeing them takes a lot of patience.*

quently climbs low shrubs and gives alarm calls from its burrow entrance; these alarm calls follow you as you walk through the colony. But in some years it can be virtually impossible to find. Occurs at high elevations of Wrangell-St. Elias NP, along Council Hwy. (look near "Trains to Nowhere" exhibit), and in many places along Dalton Hwy. (all in AK). A distinctive form with orange-tinted fur, often considered a full species called **INSULAR VOLE** (*M. insularis* or *M. abbreviatus*), occurs on St. Matthew and Hall Is. (AK) in coastal meadows.

WATER VOLE (*Microtus richardsoni*). Uncommon and local in the nw. U.S.; rare in UT and the Canadian Rockies. Inhabits moist streamside meadows at or above the timberline, marshes, grassy lakeshores, and pond edges. Mostly nocturnal and difficult to find, but sometimes can be seen walking on thin ice during the first freezes of the year. This large vole makes wide, well-worn runways along the water edge, and also "grooming spots"—little patches of bare ground on the shore, used for grooming before and after swimming. Look for it along small streams in National Bison Range (MT), in grassy wetlands in Kootenai NWR (ID), along montane streams in Grand Teton NP (WY), around small ponds near Trout Lake in Yellowstone NP (WY), above the timberline on the eastern slope of Mt. Rainier (WA), and in wet meadows on the upper slopes in Crater Lake NP (OR).

YELLOW-CHEEKED VOLE (*Microtus xanthognathus*). Also known as **TAIGA VOLE**, this uncommon rodent lives in dense colonies in sphagnum bogs, muskegs, very wet forest edges, recent burns, and clearcuts of AK and n. Canada. Colonies are ephemeral and shift locations frequently; finding them can take a lot of time, but once you find an active colony, you can usually see a vole after less than an hour of waiting. Occasionally these voles give high-pitched alarm calls as you approach their burrows. I've seen them along the southern part of Dalton Hwy. (AK), in forests near Churchill (MB), and along Hwy. 5 in the northern part of Wood Buffalo NP (NT).

ROUND-TAILED MUSKRAT (*Neofiber alleni*). Occurs in marshes, seasonally flooded grasslands, sugar cane fields, and Carolina bay–type lakes in FL and extreme se. GA. Mostly nocturnal, it can be seen by watching groups of its lodges. Look for it along the eastern side of Shark Valley Loop Rd. in Everglades NP, in Paynes Prairie PR SP, in Lake Woodruff NWR, in Mike Roess Gold Head Branch SP, along Flying Cow Rd. in West Palm Beach (all in FL), and around the observation tower in Grand Bay WMA (GA). Unfortunately, all these places except Everglades NP and Flying Cow Rd. do not normally allow nighttime access, so you are limited to the time around sunset and sunrise (unless you are able to obtain a special permit from the management). Animals in the s. part of the range are much darker than those in the n.

COMMON MUSKRAT (*Ondatra zibethicus*). Occurs in various kinds of wetlands throughout much of N. America. Usually easy to see at dusk, when it is more likely to swim across open water. Common, for example, in irrigation ditches in Grizzly Island WLA (CA), in tundra lakes in the Ogilvie Mts. (YT), along Moose-Wilson Rd. in Grand Teton NP (WY), in Lower Klamath NWR (CA), in Hot Creek SWA (CO), in Muscatatuck NWR (IN), in Sandhill WLA (WI), in Horicon NWR (WI), in Big Wall Lake NA (IA), in Potter Marsh near Anchorage (AK), and in almost every shallow, well-vegetated body of water in the East (except FL and much of TX). Lives in some cities; try Jamaica Bay in the Brooklyn part of Gateway NRA (NY). A dark-colored subspecies lives on Newfoundland. A smaller race inhabits the Gulf Coast from Freeport (TX) to Mobile Bay (AL); it is common in many wetlands along Hwy. 82 (LA).

WESTERN RED-BACKED VOLE (*Myodes californicus*). Common in coastal forests and the Cascades in OR and n. CA. Look for it in moist, dense old-growth forests with lots of large logs (but not in redwood groves), and also in unburned clearcut areas, for example, in Salt Point SP (CA), Trinity Alps (CA), and Alfred A. Loeb SP (OR). The dull-colored inland form can be found in fir forests in Crater Lake NP (OR) and Lassen Volcanic NP (CA).

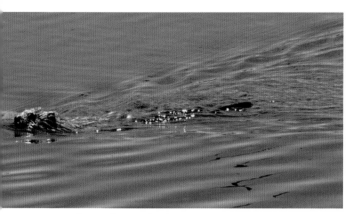

Common Muskrat is often mistaken for American Beaver. Note that it moves its tail sideways, not up and down, while swimming.

SOUTHERN RED-BACKED VOLE (*Myodes gapperi*). Common in a wide variety of wooded habitats in Canada and the ne. U.S., but limited to moist coniferous forests in the Rockies, the Cascades, and the Appalachians. Often lives in unused cabins, abandoned buildings, old chimneys remaining from burned houses in the forest, in rocky outcrops, and under large mossy logs or stumps. This brave little vole is somewhat mouselike in behavior; despite being the favorite prey of numerous predators, it often runs around in broad daylight and doesn't stay under cover as much as most other voles. A small, short-tailed rodent seen crossing a forest trail in New England or almost anywhere in Canada is most likely this species. Easy to see in Shenandoah NP (VA), in Blackwater Falls SP (WV), in Pilot Knob SP (IA), along Red Maple Swamp Trail in Trustom Pond NWR (RI), in Esther Currier WMA (NH), in Voyageurs NP (MN), in Kouchibouguac NP (NB), in Kejimkujik NP (NS), in Algonquin PP (ON), in Riding Mt. NP (MB), in Prince Albert NP (SK), and in almost any forest in NS. Introduced to Newfoundland where it's common around Little Grand Lake. Dark-colored individuals are common in the northeastern part of the range, particularly along Trans-Taiga Rd. (QC). Populations occurring from the Rockies westward are genetically distinct and might represent a separate species, closely related to Western Red-backed Vole. These voles can be seen, for example, in moist forests in Yellowstone NP (WY), along trails in Cherry Springs NA (ID), at Mt. Graham (AZ), on forested slopes in and around Valles Caldera NPR (NM), and in all NPs in the Rockies of BC.

NORTHERN RED-BACKED VOLE (*Myodes rutilus*). The most common rodent in spruce forests of AK, NU, and nw. Canada, especially in old

SOUTHERN BOG LEMMING (*Synaptomys cooperi*). Locally common in lush meadows, cool moist forests, prairie marshes, and sometimes sphagnum bogs in se. Canada, ne. U.S., and parts of the Midwest. Very difficult to trap. Present, for example, in Beaudry PP (MB), Ninigret NWR (RI), Swan Lake NWR (MO), and Bernheim Forest (KY). Relatively easy to see in appropriate habitats in NS, at the edges of small ponds in Bruce Peninsula NP (ON), along Spruce Bog Trail in Algonquin PP (ON), and along Eastmain Rd. (QC). Rare and local in the southwestern part of the range, where many populations have recently disappeared; still present on marshy shores of many prairie ponds in s. IL.

BROWN LEMMING (*Lemmus trimucronatus*). Usually the most common rodent of mainland Arctic tundra. Widespread in wet tundras and in large tundralike bogs in boreal forest in AK and n. Canada; locally also in the Canadian Rockies. Usually common around Inuvik (YT), and reportedly around Pond Inlet (NU) and other towns of s. NU. Western populations are genetically distinct and might represent a separate species, **BLACK-FOOTED LEMMING** (*L. nigripes*); this form is abundant in some years around Nome, Barrow, and along the northern part of Dalton Hwy.; it also occurs on the Alaska Peninsula and St. George I. (all in AK).

In years of low population density, Brown Lemmings are timid and difficult to find.

NEARCTIC COLLARED LEMMING (*Dicrostonyx groenlandicus*). Widespread in dry upland tundras of n. AK, n. Canada, and n. Greenland, especially in rocky areas. It is the only small mammal capable of surviving on the northernmost Arctic islands. In winter it migrates to wetter areas and can be very difficult to see, which is a pity because it changes into a gorgeous white coat. Abundant in some years around Barrow (AK), Inuvik (YT), and along the northern part of Dalton Hwy. (AK). Reportedly common on St. Lawrence I. (AK), around Pond Inlet (NU), in Ellesmere Island NP (NU), and in Northeast Greenland NP. Three southernmost subspecies are sometimes considered separate species: (1) **OGILVIE MOUNTAINS COLLARED LEMMING** (*D. nunatak-*

In a "lemming year," Brown Lemmings become fearless or even aggressive.

ensis), which is believed to be endemic to the Ogilvie Mts. (YT) and can be easily seen in dry, sparsely vegetated alpine tundra at mile 160 (km 259) of Dempster Hwy.; (2) **NELSON'S COLLARED LEMMING** (*D. nelsoni*), which is common in some years in dry upland tundras of w. AK, for example, along Teller Hwy. near Nome; and (3) **UNALASKA COLLARED LEMMING** (*D. unalascensis*), endemic to Umnak and Unalaska Is. (AK), where it is reportedly common in upland tundras. The latter is the only collared lemming that doesn't turn white in winter.

RICHARDSON'S COLLARED LEMMING (*Dicrostonyx richardsoni*). Occurs in upland tundras and in large dry openings among boreal forest from Hudson Bay to Great Slave Lake. Look for it around rocky outcrops in the vicinity of Churchill (MB). Reportedly common in the southern part of Thelon WS (NT/NU).

LABRADOR COLLARED LEMMING (*Dicrostonyx hudsonius*). Also called **UNGAVA COLLARED LEMMING**, this little-known rodent inhabits upland tundras, alpine meadows, and rocky shores of Labrador Peninsula and extreme n. QC. The only easily accessible places to see it are the high pass on Trans-Labrador Hwy. near Fermont (QC) and barren hilltops in the vicinity of Longue Pointe and Chisasibi (both in QC). You can also get to its coastal habitats by taking a ferry to any coastal village north of Goose Bay (NL).

OTHER MAMMALS

The remaining species are not related to each other, but they don't fit into any other grouping.

VIRGINIA OPOSSUM (*Didelphis virginiana*). The only marsupial in N. America, it is common in much of the e. U.S. and is gradually spreading into extreme s. Canada. Introduced to the Pacific Coast, where it occurs north to Vancouver (BC). Particularly abundant in the Southeast and in coastal CA. Animals at the northern edge of the range have the longest fur and are the most photogenic, except when they have frostbitten ears, toes, and tails after a harsh winter. Occurs in woodlands, overgrown fields, and towns. Reliable places to see opossums during night drives include Loop Rd. in Big Cypress NPR (FL), Hwy. 1 north and south of Salt Point (CA), roads in the vicinity of Cedar Point NWR (OH), Shenandoah NP (VA), and the access road to White Lake WCA (LA). Common in many towns, such as Baton Rouge (LA), Homestead (FL), and Berkeley (CA). Often visits bird feeders at night; in s. ON sometimes does so on winter days as well.

Baby opossums can often be rescued from the pouches of road-killed females. Wildlife rehabilitation centers use knit "pouches" to raise them.

NINE-BANDED ARMADILLO (*Dasypus novemcinctus*). An invasive Latin American species, known in the U.S. since the late nineteenth century. Abundant at night in woodlands, overgrown fields, and large city parks of TX, OK, AR, and the Gulf Coast; less common in KS and s. FL. Very noisy when foraging; regularly wanders into campgrounds and often can be seen on roadsides at dusk. Can be tame and easy to follow. Sites with particularly high density include Highlands Hammock SP (FL), Atchafalaya NWR (LA), Bayou Sauvage NWR (LA), as well as Palmetto and Lost Maples SNAs (TX).

MAN (*Homo sapiens*). An invasive Old World species, particularly abundant along the coasts and in the e. U.S., often in enormous colonies. The largest colony in N. America is in and around New York City (NY). Very conspicuous and noisy. Obtaining a good view is difficult because of clothing habits; naked specimens can be seen at certain beaches, mostly in FL, CA, and HI. In more remote areas its presence

Nine-banded Armadillo invaded the U.S. from Mexico in the nineteenth century. It seems to be filling the niche of a larger species (known as Beautiful Armadillo, Dasypus bellus) that was once widespread in the Southeast, but was probably hunted out by Native Americans.

can be detected by piles of garbage, although this behavior has recently become less common in N. America than in most other parts of the world. Subspecific systematics extremely controversial; 60 or more distinctive races originating from at least 10 waves of invasion can be identified visually in the field.

CARIBBEAN MANATEE (*Trichechus manatus*). Common in lagoons and shallow channels along FL coast; often very tame and playful, and will approach people and enjoy being scratched. In summer, easy to see at the manatee viewpoint in Merritt Island NWR. At Canaveral NSS you can sometimes swim with manatees in summer and early fall. In winter you can swim and snorkel with manatees in Crystal River, either on a tour or by yourself. Note that Florida law prohibits putting both hands on the same manatee (this is considered "restraint"), chasing manatees, or approaching them in the water (it's okay for them to approach you). A few manatees are often present at any time of year in

Caribbean Manatees in the clear waters of Crystal River, Florida.

Flamingo harbor in Everglades NP. If you don't mind less natural settings, you can see a lot of manatees in winter at the Big Bend Power Plant Manatee Viewing Center in Tampa, and at the Manatee Observation and Education Center in Ft. Pierce. Another good place in winter is Blue Spring SP. In summer, Florida manatees occasionally reach all Atlantic Coast states, plus NS, MS, and LA, and can be encountered at sea far from shore. The few animals recorded in TX probably belonged to the West Indian subspecies. Suffers high mortality from boat collisions and during cold winters; listed as endangered in the U.S.

SEWELLEL (*Aplodontia rufa*). This ancient rodent (also known as **MOUNTAIN BEAVER**) is locally common and sometimes considered a pest in the Pacific Northwest; rare in the Sierra Nevada. Occurs in forests, meadows, gardens, and brushy areas, often near water. In some places the ground can be so riddled with Sewellel burrows that it's difficult to walk on. Spends little time aboveground, particularly in winter. To see it, find a place with particularly high burrow density and watch the entrances for a few hours. Expect the animals to be outside for a few minutes in the first two hours after sunrise and the last two hours before sunset. Probably the best place to try is Lolo Pass near Mt. Hood (OR). Another good place is south of Irish Beach (CA) at the end of a closed road (you have to walk for a mile) called Alder Creek Beach Rd.; the colony is on the slope to the left from the terminal turnaround. Isolated subspecies inhabit Point Arena (CA) and the summit plateau of Point Reyes Peninsula (CA). An unusual desert population lives on the shores of Mono Lake (CA). The Point Arena subspecies is listed as endangered in the U.S.

AMERICAN BEAVER (*Castor canadensis*). Widespread along rivers, in shallow lakes, and in forest swamps throughout much of N. America, but local and less common in the South; abundant in much of AK and Canada. Usually builds conspicuous lodges, dams, and canals for transporting trees to the water, but in some places (particularly along large rivers) it lives in burrows, feeds on shrubs, and leaves little sign of its presence. Can be seen only when the water is not frozen; in late fall you can sometimes hear Beavers breaking thin ice to create open water channels. Beavers in some areas can be very tame, and will occasionally approach people standing still. Such places include Tallgrass Prairie NPR (KS) and Great Dismal Swamp NWR (VA/NC). Mostly crepuscular and nocturnal, American Beaver can be observed in daylight in the northern part of its range in June and July. Watch Beavers building dams and lodges at Horseshoe Lake Trail in Denali NP (AK), Kluane NP (YT), and small lakes around Nome (AK). Easy to see also in Coosawattee WMA (GA), in Hoit Road Marsh WMA (NH), in Esther Currier WMA (NH), along Pit River (CA), in Dead Horse Ranch SP (AZ), in Caddo Lake SP (TX), in Niobrara SP (NE), in Port-Daniel WR (NB), in the vicinity of Fermont (QC), in Mt. St. Helens NMT (WA), in Okmulgee WMA (OK), in Big Oaks NWR (IN), in Yeoman Park near

Eagle (CO), in Arapaho NF (CO), along Moose-Wilson Rd. in Grand Teton NP (WY), in Silver Creek Preserve (ID), in Sax-Zim Bog (MN), in Erie NWR (PA), in Fisherville Brook WR (RI), and in Forillon NP (QC). Lives in some cities, for example, in downtown Knoxville (TN), Albuquerque (NM), and Denver (CO). Scuba diving and snorkeling with Beavers is sometimes possible in the clear, warm waters of Brandy Branch Reservoir (TX). A beautiful golden-tinted race occurs in the Central Valley of CA; look for it in Caswell Memorial SP and under the footbridge across the parking lot from Amtrak station in downtown Martinez. The largest Beaver dams in the world are found in and around Wood Buffalo NP (NT/AB); there is also a very long dam in Merchants Millpond SP (NC).

Coypu is one of the most destructive introduced pests in the Southeast.

COYPU (*Myocastor coypus*). A native of S. America; introduced to numerous locations in the U.S., where it's commonly known as **NUTRIA** (which means "otter" in Spanish). Widespread and abundant in freshwater and brackish habitats along the Gulf Coast; locally common in the Pacific Northwest. Easy to see at dusk, for example, in Atchafalaya NWR (LA), Murphree WMA (TX), Sea Rim SP (TX), Mingo NWR (MO), and Goose Creek SP (NC), as well as at Capitol Lake in downtown Baton Rouge (LA).

NORTHERN PORCUPINE (*Erethizon dorsatum*). This charming rodent is common in much of AK and Canada, but uncommon and local in the ne. and w. U.S.; inhabits woodlands and riparian corridors. Nocturnal and crepuscular in the southern part of its range; less so in the North. Often seen on roadsides (especially in spring and early summer) during long drives along Trans-Labrador Hwy. (QC/NL), Alaska Hwy. (BC/YT/AK), and Glacier Hwy. (BC/YT); less often in Yellowstone NP (WY/MT/ID) and in Wood Buffalo NP (NT/AB). Occasionally seen sleeping in treetops during the day; good places to find Porcupines this way are cottonwood groves on the lands of the Blackfeet Tribe (MT), in Lostwood NWR (ND), in Warner Wetlands (OR), in Lovelock Valley (NV), in Lake Meredith NRA (TX), and around Waunita Hot Springs (CO). Occurs in tundra shrubs along Denali Hwy. (AK). Abundant in Riding Mt. NP (MB), Forillon NP (QC), and in Kinkaid Park in Anchorage (AK); reportedly common in Allegany SP (NY). Look for porcupines crossing Pumice Desert in Crater Lake NP (OR) at dusk.

PART IV.
FURTHER INFORMATION
CHECKLIST
INDEXES

FURTHER
INFORMATION

The forum and the collection of trip reports at Jon Hall's website *mammalwatching.com* is a great place to find information about mammal watching on all continents. Don't forget to send Jon your own trip reports to keep this useful resource growing. The "Birding News" page at the American Birding Association website *birding.aba .org* has links to regional birdwatching forums, where news of upcoming pelagic trips are often posted and requests for information about local specifics can be made. In particular, check out the Seabird News listserve, *birding.aba.org/maillist/SEA*. A useful resource for news on marine mammals is the MARMAM listserve, *https://lists. uvic.ca/mailman/listinfo/marmam.*

Very few good field guides for North American mammals exist, and the information they contain is often incomplete and contradictory, so it's better to have more than one. Be sure to get the latest editions, since the taxonomy is constantly changing and new species are still being described, even in well-known parts of the continent. I would particularly recommend purchasing *Peterson Field Guide to Mammals of North America* by Reid (2006) and *Mammals of North America* by Kays and Wilson (second edition, 2009). For slightly more in-depth knowledge, try *The Smithsonian Book of North American Mammals* by Wilson and Ruff (1999) and *The Natural History of Canadian Mammals* by Naughton (2012). By far the best guide to marine mammals is *Marine Mammals of the World* by Jefferson, Webber, and Pitman (2008). *Squirrels of the World* by Thorington et al. (2012) is also pretty good. *Handbook of Mammals of the World*, a popular eight-volume series being published by Lynx Edicions since 2008, is a one-stop source of information on all mammalian species, beautifully illustrated. Unfortunately, it is very expensive, suffers from poor editorial standards, and contains lots of errors; besides, some of its authors adhere to extreme minority views on taxonomy, so everything you learn from it has to be rechecked through other sources. IUCN website *iucnredlist.org* has basic data on natural history and conservation status of world's mammals, complete with distribution maps. The Smithsonian has an online database for North American mammals, *www.mnh.si.edu/mna/*. A more specialist-oriented database is the online version of the third edition of *Walker's Mammals of the World* by Wilson and Reeder (2005), available at *www.departments.bucknell.edu/biology/resources/msw3/*.

Among the numerous manuals on tracking and track identification, good books include *Mammal Tracks & Sign: A Guide to North American Species* by Elbroch (2003) and *A Field Guide to Mammal Tracking in North America* by Halfpenny and Biesiot (1986).

Of course, field guides are just very brief summaries; you'd have to read a lot more to become a real expert on wild mammals. There are plenty of books about mammals of various geographical regions, although usually you have to go through a lot of them to glean some useful information about finding particular species. A little-known but useful series, Wildlife Viewing Guides, includes numerous volumes devoted to many states and a few Canadian provinces. Sites mentioned in these books are usually marked with special road signs (brown squares with white binoculars). The California and Colorado volumes are particularly good. It is difficult to find a complete catalog, but try *www.watchablewildlife.org*, or look for titles such as *Alberta Wildlife Viewing Guide* and the like on Amazon.com. For Alaska, check out the very detailed "Wildlife Viewing" section on Alaska Fish and Game Department's website *www.adfg.alaska.gov*. For every U.S. state and every Canadian province, there should exist a huge volume titled *Mammals of . . .* (name of the state/territory)—look for them in the library of your local university (usually you have to be a student or a faculty member to check books out, but you can read them in the library, or scan/photograph the pages you need). There are also countless books focused on particular groups of mammals.

And one last notice. This book is the first of its kind, and it certainly isn't flawless. Certain regions and species are covered better than others; some information may be outdated; many interesting locations are not mentioned at all. You can help make future editions better by sending corrections, updates, and suggestions to the author at *dinets@gmail.com*.

CHECKLIST

Shrews and Moles

☐ Northern Short-tailed Shrew
(*Blarina brevicauda*)

☐ Southern Short-tailed Shrew
(*B. carolinensis*)

  ☐ Everglades Short-tailed Shrew
  (*B. peninsulae*)

  ☐ Sherman's Short-tailed Shrew
  (*B. shermani*)

☐ Elliot's Short-tailed Shrew
(*B. hylophaga*)

☐ Least Shrew (*Cryptotis parva*)

☐ Smoky Shrew (*Sorex fumeus*)

☐ Pygmy Shrew (*S. hoyi*)

☐ Dwarf Shrew (*S. nanus*)

☐ Inyo Shrew (*S. tenellus*)

☐ Ornate Shrew (*S. ornatus*)

☐ Rock Shrew (*Sorex dispar*)

  ☐ Gaspé Shrew (*S. gaspensis*)

☐ American Water Shrew
(*S. palustris*)

  ☐ Cordilleran Water Shrew
  (*S. navigator*)

  ☐ Boreal Water Shrew (*S. palustris*)

☐ Eastern Water Shrew (*S. albibarbis*)

☐ Glacier Bay Water Shrew
(*S. alaskanus*)

☐ Marsh Shrew (*S. bendirii*)

☐ Baird's Shrew (*S. bairdi*)

☐ Pacific Shrew (*S. pacificus*)

☐ Fog Shrew (*S. sonomae*)

☐ Montane Shrew (*S. monticolus*)

  ☐ New Mexican Shrew
  (*S. neomexicanus*)

☐ Vagrant Shrew (*S. vagrans*)

☐ Masked Shrew (*S. cinereus*)

☐ Maryland Shrew (*S. fontinalis*)

  ☐ St. Lawrence Island Shrew
  (*S. jacksoni*)

☐ Barren Ground Shrew
(*S. ugyunak*)

  ☐ Pribilof Islands Shrew
  (*S. pribilofensis*)

☐ Prairie Shrew (*S. haydeni*)

☐ Preble's Shrew (*S. preblei*)

☐ Mount Lyell Shrew (*S. lyelli*)

☐ Olympic Shrew (*S. rohweri*)

- [ ] Southeastern Shrew (*S. longirostris*)
- [ ] Eurasian Least Shrew (*S. minutissimus*)
- [ ] Arctic Shrew (*S. arcticus*)
- [ ] Maritime Shrew (*S. maritimensis*)
- [ ] Tundra Shrew (*S. tundrensis*)
- [ ] Arizona Shrew (*S. arizonae*)
- [ ] Merriam's Shrew (*S. merriami*)
- [ ] Trowbridge's Shrew (*S. trowbridgii*)
- [ ] Crawford's Desert Shrew (*Notiosorex crawfordi*)
- [ ] Cockrum's Desert Shrew (*N. cockrumi*)
- [ ] American Shrew Mole (*Neurotrichus gibbsi*)
- [ ] Star-nosed Mole (*Condylura cristata*)
- [ ] Broad-footed Mole (*Scapanus latimanus*)
- [ ] Coast Mole (*S. orarius*)
- [ ] Townsend's Mole (*S. townsendii*)
- [ ] Hairy-tailed Mole (*Parascalops breweri*)
- [ ] Eastern Mole (*Scalopus aquaticus*)

Bats

- [ ] Ghost-faced Bat (*Mormoops megalophylla*)
- [ ] California Leaf-nosed Bat (*Macrotus californicus*)
- [ ] Jamaican Fruit-eating Bat (*Artibeus jamaicensis*)
- [ ] Cuban Flower Bat (*Phyllonycteris poeyi*)
- [ ] Buffy Flower Bat (*Erophylla sezekorni*)
- [ ] Cuban Fig-eating Bat (*Phyllops falcatus*)
- [ ] Mexican Long-tongued Bat (*Choeronycteris mexicana*)
- [ ] Lesser Long-nosed Bat (*Leptonycteris yerbabuenae*)
- [ ] Greater Long-nosed Bat (*L. nivalis*)
- [ ] Little Brown Myotis (*Myotis lucifugus*)
- [ ] Fringed Myotis (*M. thysanodes*)
- [ ] Long-eared Myotis (*M. evotis*)
- [ ] Arizona Myotis (*M. occultus*)
- [ ] Long-legged Myotis (*M. volans*)
- [ ] Indiana Myotis (*M. sodalis*)
- [ ] Southwestern Myotis (*M. auriculus*)
- [ ] Northern Myotis (*M. septentrionalis*)
- [ ] California Myotis (*M. californicus*)
- [ ] Eastern Small-footed Myotis (*M. leibii*)
- [ ] Western Small-footed Myotis (*M. ciliolabrum*)
- [ ] Yuma Myotis (*M. yumanensis*)
- [ ] Cave Myotis (*M. velifer*)
- [ ] Gray Myotis (*M. grisescens*)
- [ ] Southeastern Myotis (*M. austroriparius*)

- ☐ Seminole Bat (*Lasiurus seminolus*)
- ☐ Western Red Bat (*L. blossevillii*)
- ☐ Eastern Red Bat (*L. borealis*)
- ☐ Hoary Bat (*L. cinereus*)
- ☐ Southern Yellow Bat (*L. ega*)
- ☐ Western Yellow Bat (*L. xanthinus*)
- ☐ Northern Yellow Bat (*L. intermedius*)
- ☐ Silver-haired Bat (*Lasionycteris noctivagans*)
- ☐ Canyon Bat (*Parastrellus hesperus*)
- ☐ Tricolored Bat (*Perimyotis subflavus*)
- ☐ Big Brown Bat (*Eptesicus fuscus*)
- ☐ Evening Bat (*Nycticeius hymeralis*)
- ☐ Spotted Bat (*Euderma maculatum*)
- ☐ Allen's Big-eared Bat (*Idionycteris phyllotis*)
- ☐ Rafinesque's Big-eared Bat (*Corynorhinus rafinesquii*)
- ☐ Townsend's Big-eared Bat (*C. townsendii*)
- ☐ Mexican Big-eared Bat (*C. mexicanus*)
- ☐ Pallid Bat (*Antrozous pallidus*)
- ☐ Mexican Free-tailed Bat (*Tadarida brasiliensis*)
- ☐ Pocketed Free-tailed Bat (*Nyctinomops femorosaccus*)
- ☐ Big Free-tailed Bat (*N. macrotis*)

- ☐ Florida Bonneted Bat (*Eumops floridanus*)
- ☐ Western Bonneted Bat (*E. perotis*)
- ☐ Underwood's Bonneted Bat (*E. underwoodi*)
- ☐ Velvety Free-tailed Bat (*Molossus molossus*)

Carnivores

- ☐ Gray Wolf (*Canis lupus*)
- ☐ Red Wolf (*C. rufus*)
  - ☐ Algonquin Wolf (*C. lycaon*)
- ☐ Coyote (*C. latrans*)
- ☐ Red Fox (*Vulpes vulpes*)
- ☐ Swift Fox (*V. velox*)
  - ☐ Kit Fox (*V. macrotis*)
- ☐ Arctic Fox (*V. lagopus*)
- ☐ Gray Fox (*Urocyon cinereoargenteus*)
- ☐ Island Fox (*U. littoralis*)
- ☐ American Black Bear (*Ursus americanus*)
- ☐ Brown Bear (*U. arctos*)
- ☐ Polar Bear (*U. maritimus*)
- ☐ Ringtail (*Bassariscus astutus*)
- ☐ Northern Raccoon (*Procyon lotor*)
- ☐ White-nosed Coati (*Nasua narica*)
- ☐ Wolverine (*Gulo gulo*)
- ☐ American Marten (*Martes americana*)
  - ☐ Pacific Marten (*M. caurina*)
- ☐ Fisher (*M. pennanti*)

- ☐ Least Weasel (*Mustela nivalis*)
- ☐ Ermine (*M. erminea*)
- ☐ Long-tailed Weasel (*M. frenata*)
- ☐ Black-footed Ferret (*M. nigripes*)
- ☐ American Mink (*M. vison*)
- ☐ American Badger (*Taxidea taxus*)
- ☐ Northern River Otter (*Lontra canadensis*)
- ☐ Sea Otter (*Enhydra lutris*)
- ☐ Western Spotted Skunk (*Spilogale gracilis*)
- ☐ Eastern Spotted Skunk (*S. putorius*)
- ☐ Striped Skunk (*Mephitis mephitis*)
- ☐ Hooded Skunk (*M. macroura*)
- ☐ White-backed Hog-nosed Skunk (*Conepatus leuconotus*)
- ☐ Cougar (*Puma concolor*)
- ☐ Jaguarundi (*P. yagouaroundi*)
- ☐ Ocelot (*Leopardus pardalis*)
- ☐ Canada Lynx (*Lynx canadensis*)
- ☐ Bobcat (*L. rufus*)
- ☐ Jaguar (*Panthera onca*)

Pinnipeds

- ☐ Northern Fur Seal (*Callorhinus ursinus*)
- ☐ Guadalupe Fur Seal (*Arctocephalus townsendi*)
- ☐ Steller's Sea Lion (*Eumetopias jubatus*)
- ☐ California Sea Lion (*Zalophus californianus*)
- ☐ Walrus (*Odobenus rosmarus*)
- ☐ Harbor Seal (*Phoca vitulina*)
- ☐ Spotted Seal (*P. largha*)
- ☐ Ringed Seal (*P. hispida*)
- ☐ Gray Seal (*P. grypa*)
- ☐ Harp Seal (*Pagophilus groenlandicus*)
- ☐ Ribbon Seal (*Histriophoca fasciata*)
- ☐ Bearded Seal (*Erignathus barbatus*)
- ☐ Hooded Seal (*Cystophora cristata*)
- ☐ Hawaiian Monk Seal (*Monachus schauinslandi*)
- ☐ Northern Elephant Seal (*Mirounga angustirostris*)

Cetaceans

- ☐ Gray Whale (*Eschrichtius robustus*)
- ☐ Northern Minke Whale (*Balaenoptera acutorostrata*)
- ☐ Bryde's Whale (*B. brydei*)
- ☐ Sei Whale (*B. borealis*)
- ☐ Fin Whale (*B. physalus*)
- ☐ Blue Whale (*B. musculus*)
- ☐ Humpback Whale (*Megaptera novaeangliae*)
- ☐ Northern Right Whale (*Eubalaena glacialis*)
  - ☐ North Pacific Right Whale (*E. japonica*)

- [ ] Bowhead Whale (*Balaena mysticetus*)
- [ ] Beluga (*Delphinapterus leucas*)
- [ ] Narwhal (*Monodon monoceros*)
- [ ] Rough-toothed Dolphin (*Steno bredanensis*)
- [ ] Common Bottlenose Dolphin (*Tursiops truncatus*)
- [ ] Pantropical Spotted Dolphin (*Stenella attenuata*)
- [ ] Atlantic Spotted Dolphin (*S. frontalis*)
- [ ] Clymene Dolphin (*S. clymene*)
- [ ] Spinner Dolphin (*S. longirostris*)
- [ ] Striped Dolphin (*S. coeruleoalba*)
- [ ] Long-beaked Common Dolphin (*Delphinus capensis*)
- [ ] Short-beaked Common Dolphin (*D. delphis*)
- [ ] Fraser's Dolphin (*Lagenodelphis hosei*)
- [ ] Atlantic White-sided Dolphin (*Lagenorhynchus acutus*)
- [ ] White-beaked Dolphin (*L. albirostris*)
- [ ] Pacific White-sided Dolphin (*L. obliquidens*)
- [ ] Northern Right Whale Dolphin (*Lissodelphis borealis*)
- [ ] Risso's Dolphin (*Grampus griseus*)
- [ ] Long-finned Pilot Whale (*Globicephala melas*)
- [ ] Short-finned Pilot Whale (*G. macrorhynchus*)
- [ ] Melon-headed Whale (*Peponocephala electra*)
- [ ] Pygmy Killer Whale (*Feresa attenuata*)
- [ ] False Killer Whale (*Pseudorca crassidens*)
- [ ] Killer Whale (*Orcinus orca*)
- [ ] Harbor Porpoise (*Phocoena phocoena*)
- [ ] Dall's Porpoise (*P. dalli*)
- [ ] Baird's Beaked Whale (*Berardius bairdi*)
- [ ] Northern Bottlenose Whale (*Hyperoodon ampullatus*)
- [ ] Cuvier's Beaked Whale (*Ziphius cavirostris*)
- [ ] Sowerby's Beaked Whale (*Mesoplodon bidens*)
- [ ] Gervais's Beaked Whale (*M. europaeus*)
- [ ] True's Beaked Whale (*M. mirus*)
- [ ] Blainville's Beaked Whale (*M. densirostris*)
- [ ] Stejneger's Beaked Whale (*M. stejnegeri*)
- [ ] Hubbs's Beaked Whale (*M. carlhubbsi*)
- [ ] Perrin's Beaked Whale (*M. perrini*)
- [ ] Ginkgo-toothed Beaked Whale (*M. ginkgodens*)
- [ ] Pygmy Beaked Whale (*M. peruvianus*)

- [ ] Tropical Beaked Whale (*M. pacificus*)
- [ ] Pygmy Sperm Whale (*Kogia breviceps*)
- [ ] Dwarf Sperm Whale (*K. sima*)
- [ ] Sperm Whale (*Physeter macrocephalus*)

Ungulates

- [ ] Collared Peccary (*Pecari tajacu*)
- [ ] Elk (*Cervus canadensis*)
- [ ] Mule Deer (*Odocoileus hemionus*)
- [ ] White-tailed Deer (*O. virginianus*)
- [ ] Moose (*Alces alces*)
- [ ] Caribou (*Rangifer tarandus*)
- [ ] Pronghorn (*Antilocapra americana*)
- [ ] Mountain Goat (*Oreamnos americanus*)
- [ ] Muskox (*Ovibos moschatus*)
- [ ] Bighorn Sheep (*Ovis canadensis*)
- [ ] Dall Sheep (*O. dalli*)
- [ ] American Bison (*Bos bison*)

Lagomorphs

- [ ] Collared Pika (*Ochotona collaris*)
- [ ] American Pika (*O. princeps*)
- [ ] Pygmy Rabbit (*Brachylagus idahoensis*)
- [ ] Brush Rabbit (*Sylvilagus bachmani*)
- [ ] Eastern Cottontail (*S. floridanus*)

- [ ] Davis Mountains Cottontail (*S. robustus*)
- [ ] Manzano Mountains Cottontail (*S. cognatus*)
- [ ] New England Cottontail (*S. transitionalis*)
  - [ ] Appalachian Cottontail (*S. obscurus*)
- [ ] Desert Cottontail (*S. audubonii*)
- [ ] Mountain Cottontail (*S. nuttallii*)
- [ ] Swamp Rabbit (*S. aquaticus*)
- [ ] Marsh Rabbit (*S. palustris*)
- [ ] White-tailed Jackrabbit (*Lepus townsendii*)
- [ ] Antelope Jackrabbit (*L. alleni*)
- [ ] Black-tailed Jackrabbit (*L. californicus*)
- [ ] White-sided Jackrabbit (*L. callotis*)
- [ ] Arctic Hare (*L. arcticus*)
  - [ ] Alaska Hare (*L. othus*)
- [ ] Snowshoe Hare (*L. americanus*)

Odd Rodents

- [ ] Sewellel (*Aplodontia rufa*)
- [ ] Northern Porcupine (*Erethizon dorsatum*)
- [ ] American Beaver (*Castor canadensis*)

Sciurids

- [ ] Northern Flying Squirrel (*Glaucomys sabrinus*)

- [ ] Southern Flying Squirrel (*G. volans*)
- [ ] Pine Squirrel (*Tamiasciurus hudsonicus*)
- [ ] Abert's Squirrel (*Sciurus aberti*)
- [ ] Mexican Fox Squirrel (*S. nayaritensis*)
- [ ] Eastern Fox Squirrel (*S. niger*)
- [ ] Arizona Gray Squirrel (*S. arizonensis*)
- [ ] Eastern Gray Squirrel (*S. carolinensis*)
- [ ] Western Gray Squirrel (*S. griseus*)
- [ ] Eastern Chipmunk (*Tamias striatus*)
- [ ] Yellow-pine Chipmunk (*T. amoenus*)
- [ ] Gray-footed Chipmunk (*T. canipes*)
- [ ] Gray-collared Chipmunk (*T. cinereicollis*)
- [ ] Cliff Chipmunk (*T. dorsalis*)
- [ ] Hopi Chipmunk (*T. rufus*)
- [ ] Colorado Chipmunk (*T. quadrivittatus*)
- [ ] Uinta Chipmunk (*T. umbrinus*)
- [ ] Palmer's Chipmunk (*T. palmeri*)
- [ ] Panamint Chipmunk (*T. panamintinus*)
- [ ] Least Chipmunk (*T. minimus*)
- [ ] Alpine Chipmunk (*T. alpinus*)
- [ ] Long-eared Chipmunk (*T. quadrimaculatus*)
- [ ] Lodgepole Chipmunk (*T. speciosus*)
- [ ] California Chipmunk (*T. obscurus*)
- [ ] Merriam's Chipmunk (*T. merriami*)
- [ ] Sonoma Chipmunk (*T. sonomae*)
- [ ] Yellow-cheeked Chipmunk (*T. ochrogenys*)
- [ ] Allen's Chipmunk (*T. senex*)
- [ ] Siskiyou Chipmunk (*T. siskiyou*)
- [ ] Townsend's Chipmunk (*T. townsendii*)
- [ ] Red-tailed Chipmunk (*T. ruficaudus*)
- [ ] Harris's Antelope Squirrel (*Ammospermophilus harrisii*)
- [ ] Texas Antelope Squirrel (*A. interpres*)
- [ ] White-tailed Antelope Squirrel (*A. leucurus*)
- [ ] Nelson's Antelope Squirrel (*A. nelsoni*)
- [ ] Golden-mantled Ground Squirrel (*Callospermophilus lateralis*)
- [ ] Cascade Ground Squirrel (*C. saturatus*)
- [ ] California Ground Squirrel (*Otospermophilus beecheyi*)
- [ ] Rock Squirrel (*O. variegatus*)
- [ ] Arctic Ground Squirrel (*Urocitellus parryii*)
- [ ] Uinta Ground Squirrel (*U. armatus*)
- [ ] Belding's Ground Squirrel (*U. beldingi*)

- ☐ Idaho Ground Squirrel (*U. brunneus*)

- ☐ Piute Ground Squirrel (*U. mollis*)

- ☐ Townsend's Ground Squirrel (*U. townsendii*)

- ☐ Merriam's Ground Squirrel (*U. canus*)

- ☐ Columbian Ground Squirrel (*U. columbianus*)

- ☐ Wyoming Ground Squirrel (*U. elegans*)

- ☐ Richardson's Ground Squirrel (*U. richardsonii*)

- ☐ Washington Ground Squirrel (*U. washingtoni*)

- ☐ Thirteen-lined Ground Squirrel (*Ictidomys tridecimlineatus*)

- ☐ Mexican Ground Squirrel (*I. mexicanus*)

- ☐ Mohave Ground Squirrel (*Xerospermophilus mohavensis*)

- ☐ Spotted Ground Squirrel (*X. spilosoma*)

- ☐ Round-tailed Ground Squirrel (*X. tereticaudus*)

- ☐ Franklin's Ground Squirrel (*Poliocitellus franklinii*)

- ☐ Gunnison's Prairie Dog (*Cynomus gunnisoni*)

- ☐ Utah Prairie Dog (*C. parvidens*)

- ☐ White-tailed Prairie Dog (*C. leucurus*)

- ☐ Black-tailed Prairie Dog (*C. ludovicianus*)

- ☐ Yellow-bellied Marmot (*Marmota flaviventris*)

- ☐ Olympic Marmot (*M. olympus*)

- ☐ Vancouver Island Marmot (*M. vancouverensis*)

- ☐ Hoary Marmot (*M. caligata*)

- ☐ Alaska Marmot (*M. broweri*)

- ☐ Woodchuck (*M. monax*)

Pocket Gophers

- ☐ Botta's Pocket Gopher (*Thomomys bottae*)

- ☐ Camas Pocket Gopher (*T. bulbivorus*)

- ☐ Western Pocket Gopher (*T. mazama*)

- ☐ Mountain Pocket Gopher (*T. monticola*)

- ☐ Northern Pocket Gopher (*T. talpoides*)

- ☐ Idaho Pocket Gopher (*T. idahoensis*)

- ☐ Wyoming Pocket Gopher (*T. clusius*)

- ☐ Desert Pocket Gopher (*Geomys arenarius*)

- ☐ Attwater's Pocket Gopher (*G. attwateri*)

- ☐ Baird's Pocket Gopher (*G. breviceps*)

- ☐ Plains Pocket Gopher (*G. bursarius*)

- ☐ Jones's Pocket Gopher (*G. knoxjonesi*)

- ☐ Llano Pocket Gopher (*G. texensis*)
- ☐ Texas Pocket Gopher (*G. personatus*)
  - ☐ Strecker's Pocket Gopher (*G. streckeri*)
- ☐ Southeastern Pocket Gopher (*G. pinetis*)
- ☐ Yellow-faced Pocket Gopher (*Cratogeomys castanops*)

Heteromyids

- ☐ Mexican Spiny Pocket Mouse (*Liomys irroratus*)
- ☐ Hispid Pocket Mouse (*Chaetodipus hispidus*)
- ☐ Long-tailed Pocket Mouse (*C. formosus*)
- ☐ Rock Pocket Mouse (*C. intermedius*)
- ☐ Spiny Pocket Mouse (*C. spinatus*)
- ☐ California Pocket Mouse (*C. californicus*)
- ☐ San Diego Pocket Mouse (*C. fallax*)
- ☐ Nelson's Pocket Mouse (*C. nelsoni*)
- ☐ Desert Pocket Mouse (*C. penicillatus*)
  - ☐ Chihuahuan Pocket Mouse (*C. eremicus*)
- ☐ Bailey's Pocket Mouse (*C. baileyi*)
- ☐ Baja Pocket Mouse (*C. rudinoris*)
- ☐ Plains Pocket Mouse (*Perognathus flavescens*)
- ☐ Olive-backed Pocket Mouse (*P. fasciatus*)
- ☐ Silky Pocket Mouse (*P. flavus*)
- ☐ Merriam's Pocket Mouse (*P. merriami*)
- ☐ Little Pocket Mouse (*P. longimembris*)
- ☐ San Joaquin Pocket Mouse (*P. inornatus*)
- ☐ Arizona Pocket Mouse (*P. amplus*)
- ☐ Great Basin Pocket Mouse (*P. parvus*)
- ☐ White-eared Pocket Mouse (*P. alticolus*)
- ☐ Dark Kangaroo Mouse (*Microdipodops megacephalus*)
- ☐ Pale Kangaroo Mouse (*M. pallidus*)
- ☐ Ord's Kangaroo Rat (*Dipodomys ordii*)
- ☐ Gulf Coast Kangaroo Rat (*D. compactus*)
- ☐ Desert Kangaroo Rat (*D. deserti*)
- ☐ Banner-tailed Kangaroo Rat (*D. spectabilis*)
- ☐ Texas Kangaroo Rat (*D. elator*)
- ☐ Merriam's Kangaroo Rat (*D. merriami*)
- ☐ Chisel-toothed Kangaroo Rat (*D. microps*)
- ☐ Fresno Kangaroo Rat (*D. nitratoides*)
- ☐ California Kangaroo Rat (*D. californicus*)

- ☐ Giant Kangaroo Rat (*D. ingens*)
- ☐ Agile Kangaroo Rat (*D. agilis*)
- ☐ Dulzura Kangaroo Rat (*D. simulans*)
- ☐ Narrow-faced Kangaroo Rat (*D. venustus*)
- ☐ Heermann's Kangaroo Rat (*D. heermanni*)
- ☐ Panamint Kangaroo Rat (*D. panamintinus*)
- ☐ Stephens's Kangaroo Rat (*D. stephensi*)

Mice and Rats

- ☐ Marsh Rice Rat (*Oryzomys palustris*)
- ☐ Coues's Rice Rat (*O. couesi*)
- ☐ Eastern Harvest Mouse (*Reithrodontomys humulis*)
- ☐ Fulvous Harvest Mouse (*R. fulvescens*)
- ☐ Plains Harvest Mouse (*R. montanus*)
- ☐ Western Harvest Mouse (*R. megalotis*)
- ☐ Salt-marsh Harvest Mouse (*R. raviventris*)
- ☐ Golden Mouse (*Ochrotomys nuttalli*)
- ☐ Florida Mouse (*Peromyscus floridanus*)
- ☐ Deer Mouse (*P. maniculatus*)
- ☐ Oldfield Mouse (*P. polionotus*)
- ☐ Keen's Mouse (*P. keeni*)

- ☐ Black-eared Mouse (*P. melanotis*)
- ☐ White-footed Mouse (*P. leucopus*)
- ☐ Cotton Mouse (*P. gossypinus*)
- ☐ Cactus Mouse (*P. eremicus*)
- ☐ Northern Baja Mouse (*P. fraterculus*)
- ☐ Mesquite Mouse (*P. merriami*)
- ☐ Canyon Mouse (*P. crinitus*)
- ☐ Brush Mouse (*P. boylii*)
- ☐ Texas Mouse (*P. attwateri*)
- ☐ White-ankled Mouse (*P. pectoralis*)
- ☐ Northern Rock Mouse (*P. nasutus*)
- ☐ Osgood's Mouse (*P. gratus*)
- ☐ Piñon Mouse (*P. truei*)
- ☐ California Mouse (*P. californicus*)
- ☐ Northern Grasshopper Mouse (*Onychomys leucogaster*)
- ☐ Mearns's Grasshopper Mouse (*O. arenicola*)
- ☐ Southern Grasshopper Mouse (*O. torridus*)
- ☐ Northern Pygmy Mouse (*Baiomys taylori*)
- ☐ Eastern Woodrat (*Neotoma floridana*)
- ☐ Appalachian Woodrat (*N. magister*)
- ☐ White-throated Woodrat (*N. albigula*)
- ☐ White-toothed Woodrat (*N. leucodon*)

LOCATIONS INDEX

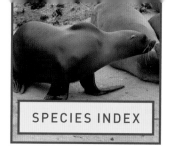

SPECIES INDEX

Index pages in **bold** refer to text graphics.

FISHES

Atlantic Coast Fishes

Freshwater Fishes

SPACE

Stars and Planets

GEOLOGY

Geology of Eastern North America

Rocks and Minerals

SEASHORE

Atlantic Seashore

Shells of the Atlantic and Gulf Coasts and the West Indies

Southeastern and Caribbean Seashores

PETERSON FLASHGUIDES®

Portable and waterproof, FlashGuides are perfect for those who want to travel light. Covering 50–100 species, with brief surveys of habit and habitat, each opens to two rows with twelve full-color, laminated panels on each side.

Atlantic Coastal Birds

Birds of the Midwest

Trees

PETERSON FIRST GUIDES®

The first books the beginning naturalist needs, whether young or old. Simplified versions of the full-size guides, they make it easy to get started in the field, and feature the most commonly seen natural life.

Astronomy

Birds

Butterflies and Moths

Caterpillars

Clouds and Weather

Fishes

Insects

Mammals

Reptiles and Amphibians

Rocks and Minerals

Seashores

Shells

Trees

Urban Wildlife

Wildflowers

PETERSON FIELD GUIDES FOR YOUNG NATURALISTS

This series is designed with young readers ages eight to twelve in mind, featuring the original artwork of the celebrated naturalist Roger Tory Peterson.

Backyard Birds

Birds of Prey

Songbirds

Butterflies

Caterpillars

PETERSON FIELD GUIDES® COLORING BOOKS®

Fun for kids ages eight to twelve, these color-your-own field guides include color stickers and are suitable for use with pencils or paint.

Birds

Butterflies

Dinosaurs

Reptiles and Amphibians

Wildflowers

Seashores

Shells

Mammals

PETERSON REFERENCE GUIDES®

Reference Guides provide in-depth information on groups of birds and topics beyond identification.

Seawatching: Eastern Waterbirds in Flight

Gulls of the Americas

Molt in North American Birds

Behavior of North American Mammals

Birding by Impression

PETERSON AUDIO GUIDES

Birding by Ear: Western

Birding by Ear: Eastern/Central

More Birding by Ear: Eastern/Central

Bird Songs: Eastern/Central

PETERSON FIELD GUIDE / *BIRD WATCHER'S DIGEST* BACKYARD BIRD GUIDES

Identifying and Feeding Birds

Hummingbirds and Butterflies

Bird Homes and Habitats

The Young Birder's Guide to Birds of North America

The New Birder's Guide to Birds of North America

DIGITAL

Apps available on the App Store for iPad, iPhone, and iPod Touch.

Peterson Birds of North America

Peterson Birds Pocket Edition

Peterson Backyard Birds

E-books

Birds of Arizona

Birds of California

Birds of Florida

Birds of Massachusetts

Birds of Minnesota

Birds of New Jersey

Birds of New York

Birds of Ohio

Birds of Pennsylvania

Birds of Texas